STEP-BY-STEP GUIDE TO
BRAKE SERVICING

STEP-BY-STEP GUIDE TO
BRAKE SERVICING

BY HERSCHEL & RAY WHITTINGTON

TAB BOOKS
Blue Ridge Summit, Pa. 17214

FIRST EDITION

THIRD PRINTING—APRIL 1980

Printed in the United States of America

Copyright © 1975 by TAB BOOKS

Hardbound Edition: International Standard Book No. 0-8306-5818-1

Paperbound Edition: International Standard Book No. 0-8306-4818-6

Library of Congress Card Number: 75-31465

Contents

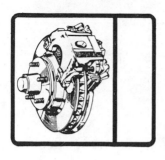

Preface

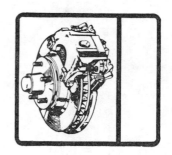

A complete brake job—linings, drums and discs turned, cylinders overhauled—costs as much as $250, depending upon the car and local labor rates. With the help of this Step-by-Step Guide you can do the job yourself and save $100 or more.

By the same token, this Guide will take you step by step through every aspect of brake work from adjusting the brakes, to rebuilding the master cylinder, to tightening the wheel bearings. All American brakes since 1960 are covered in depth, including Studebaker four-disc models and Corvette disc and drum combinations. You will learn all about power brakes, metering and proportioning valves, brake fluid, lines and hoses, and scores of other vital subjects. You will learn the correct and easy way to adjust drum and parking brakes, how to flush and bleed the hydraulic system, and dozens of the tricks of the trade that only specialists know.

The authors want to thank the folks at *Bendix Automotive Aftermarket* for their tremendous cooperation during the preparation of this book.

Herschel and Ray T. Whittington

Brake System Fundamentals

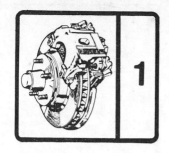

1

Hydraulic brakes were first used on the Rickenbacker car of a half-century ago. Some say the car's novel brakes were responsible for its demise. Rival manufacturers engaged in a whisper campaign to the effect that the Rickenbacker's hydraulic lines had a tendency to snap like spaghetti.

But hydraulics were the wave of the future. By 1940 all American cars used them. The older brakes depended upon rods and cables to actuate the shoes. While this system was reasonably reliable, it was subject to severe limitations as car speeds increased. It was almost impossible to keep mechanical brakes in adjustment. Inevitably the linkage to one wheel would stretch or wear more than the others. But the major problem was that stopping pressure was entirely a function of muscle and leverage.

BRAKE PRINCIPLES

A hydraulic system automatically equalizes pressure. The only adjustment required is to set the shoes the same distance from the inside diameter of the drum. And a hydraulic system is designed to give a mechanical advantage. Most manufacturers consider that 70 lb on the pedal should be the maximum required force to lock the wheels at any speed the car is capable of attaining. This figure has been revised downward with the advent of power boosters. Pressure on the shoes and pads is on the order of three tons in a panic stop.

The hydraulic system basically consists of a master cylinder, brake lines and flexible brake hoses, and slave cylinders at the wheels. Fluid is the working medium. Figure 1-1 illustrates such a system.

When the driver presses on the brake pedal several things happen. The force his leg muscles exert is multiplied mechanically by the pedal arm. As you can observe in the illustration, the *fulcrum* or pivot of the pedal is a short distance above the pushrod that goes into the power booster

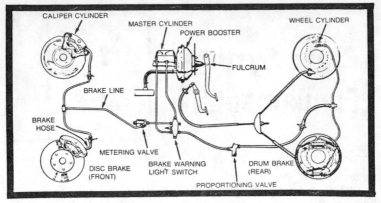

Fig. 1-1. Typical modern car brake system valves and switches.

and master cylinder assembly. The position of the fulcrum multiplies the force extended on the pedal. But the significant multiplication occurs in the hydraulic system.

The master cylinder consists of a piston and a bore with a fluid reservoir on top. The wheel cylinders contain one or two pistons and are connected directly to the brake shoes or pads (Fig. 1-2). Brake fluid does not significantly compress; pressures in the system are everywhere the same.

Why? Let us suppose that the master-cylinder piston has an area of 5 square inches. If we have 50 pounds per square inch (psi) in the system, the force acting against the face of the piston is five times that, or 250 psi. The wheel-cylinder pistons are larger than the master-cylinder piston. And there are four of them. The hydraulic system "sees" these four pistons as a single giant piston. Its total surface area may be in the neighborhood of 50 square inches. Fifty pounds per square inch acting on a surface of 50 square inches generates a force of 2500 lb. We have increased force by a factor of 10.

Of course, one doesn't get something for nothing. The gain in force is matched by a reduction in wheel-cylinder piston travel. In our example these pistons move only one tenth as far as the master-cylinder piston.

The brake mechanism consists of two friction surfaces. One is in the shape of a drum or a disc. The drawing in Fig. 1-1 illustrates a mixed system with disc brakes on the front wheels and drums on the rear. These parts are secured to the wheels and rotate with them. The disc (or *rotor* as it is more properly called) has two or four *pads* poised near it. The design shown has four. The drums each contain a pair of shoes known as the *primary* and the *secondary*. Pads and shoes are fixed to the chassis and cannot rotate.

When the system is energized, the pads and shoes, impelled by hydraulic force, make contact with the discs and drums. Friction between these parts and between the tires and the road surface causes the vehicle to slow and stop.

BRAKE FLUID

Brake fluid is perhaps the single most important component of the system because it links it all together. It is a specially blended liquid, required by law to meet certain standards, and embracing the following characteristics:

Viscosity: Free flowing at all operating temperatures.

Boiling point: Remain liquid at highest operating temperatures.

Corrosiveness: Must not attack metal or rubber parts.

Water tolerance: Must absorb and retain moisture that collects in system.

Lubricating ability: Must lubricate pistons and cups to reduce internal friction and wear.

Low freezing point: Must not freeze, even at lowest operating temperatures.

Getting the proper brake fluid is easy enough under normal circumstances; just be sure the container is inscribed with this specifications: *DOT-3*. Actually, higher performance DOT-4 fluid is available but, unless you drive like Richard Petty at Daytona, the DOT-3 is more than adequate.

With use, brake fluid tends to become contaminated with moisture, dust, and minute particles of metal, rubber, and paint from internal parts of the system. Any time you perform major work on a brake system you should flush the system:

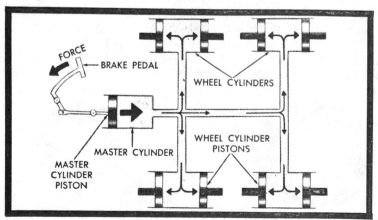

Fig. 1-2. Schematic of a typical brake system.

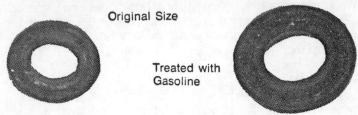

Original Size

Treated with
Gasoline

Fig. 1-3. Wheel-cylinder cup treated with gasoline.

drain out the old fluid, flush new fluid through it, then refill it
with new fluid.

On most cars, discoloration of fluid in the system is
an indication of contamination; however, according to the
June 1974 issue of *Service Digest* (Bendix), brake fluid in
the front reservoir of master cylinders on Gremlins and
Hornets with manual disc brakes has a milky appearance,
while the fluid in the rear reservoir is clean. Colors of light
yellow, pink, orange, or purple may be encountered on other
models as well. This discoloration, the report continues, is
caused by "the dissolving of specific color dye used by the
manufacturer to identify the spring in the master cylinder
bore, and in no way affects or impairs brake operation."

Keep brake fluid, whether in your brake system or stored
in a container, sealed tightly. Not only does air contain
moisture, but also almost all air carries other contaminants.
Any foreign matter is likely to render brake fluid useless. And
never, never place hydraulic fluid in a container that is not
100% free of a residual petroleum product. One drop can put
you out of business.

If you don't believe it, try the experiment illustrated in
Fig. 1-3. The "tumor" on that wheel-cylinder cup arose within
20 minutes after we applied gasoline to it. Then, to drive home
our point more dramatically, we dropped another cup into a
bowl of motor oil for 60 minutes, then took it out and
photographed it along side a new cup (Fig. 1-4).

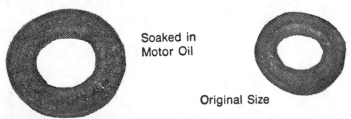

Soaked in
Motor Oil

Original Size

Fig. 1-4. Clean cup compared with identical cup soaked in motor oil for 60
minutes.

Master Cylinder and Power Booster

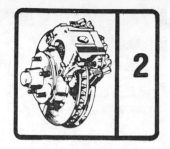

2

Yes, our cars do still use foot-operated brake pedals.

Why?

Because "we've always done it that way."

Other means of activating brakes might work better, but the pedal is a pretty good way of doing it, and we're all used to it by now. So, because it seems likely that this device is going to be with us for awhile, we might as well take a brief look at what it is, what it does, and how it does it.

Originally the pedal came into use on wagons. Its purpose was to exert more braking power on a wheel by combining leg power—which is greater than arm power—with leverage. And it's still used for much the same reason, though modern power-assisted hydraulic systems depend much less on leg power or leverage. Even so, brake pedal design has changed little down through the years. (See Figs. 2-1, 2-2, and 2-3.)

With few exceptions, the brake pedal is coupled by a clevis pushrod assembly to the heart of your hydraulic brake system, the *master cylinder*.

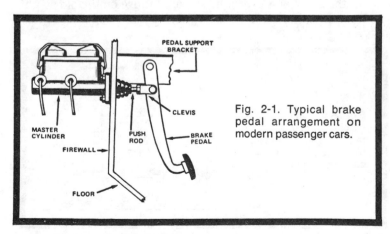

Fig. 2-1. Typical brake pedal arrangement on modern passenger cars.

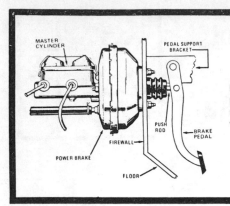

Fig. 2-2. Typical brake pedal arrangement on modern passenger cars with power brakes.

MASTER CYLINDER

There are two basic types of master cylinders in use today: the single system and the dual system.

Single-System Master Cylinder

As of production of 1968 models, all new cars sold in the U.S. were required to have *dual* hydraulic systems for safety reasons. Prior to that time, many American production cars employed what we now call a "single system," which quite naturally required only a single-system master cylinder. Such master cylinders (found on *all* American-built cars prior to 1963) are easily distinguished from dual-system master cylinders by their single reservoirs. Compare Fig. 2-4 with Fig. 2-5.

Dual-System Master Cylinder

In the event of a hydraulic system failure on the old single system, the failure was catastrophic—total. To overcome this

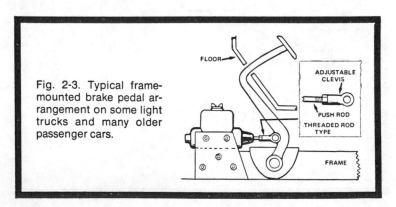

Fig. 2-3. Typical frame-mounted brake pedal arrangement on some light trucks and many older passenger cars.

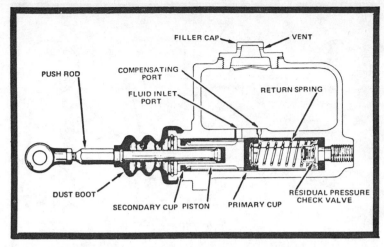

Fig. 2-4. Typical single-system master cylinder.

situation, the dual system was made mandatory. In essence, all it did was to make the front-wheel brakes a separate system from the rear-wheel brakes.

This separation of the system augmented an even greater advance in braking—use of the much more efficient disc brakes on front wheels. Essentially, a separate system of hydraulic transfer is required for disc brakes anyhow, so federal regulations and advanced engineering worked

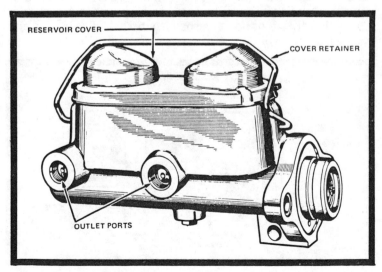

Fig. 2-5. Typical dual-system master cylinder.

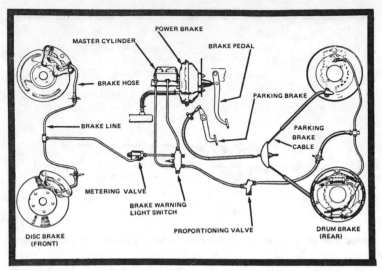

Fig. 2-6. Typical modern braking system.

hand-in-hand during the decade of the sixties to give us the greatest advance in braking systems since Chevrolet adopted servos and Chrysler went with hydraulics back in the 1920s.

Typical of the improved system of the sixties is that shown in Fig. 2-6. This system is served by a distinctive dual-system master cylinder such as that shown in Fig. 2-7. You will notice that this master cylinder has two reservoirs, though they differ in size—the larger reservoir serves the front disc brakes. This distinctive difference is always the easiest-to-spot clue as to whether or not a car has front-wheel disc brakes and rear-wheel drum brakes.

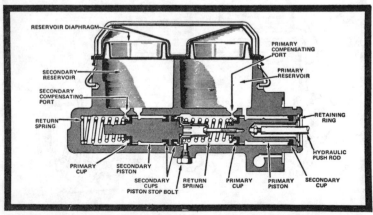

Fig. 2-7. The working parts of a typical dual-system master cylinder.

But, just as telltale is the master cylinder for a car on which all four wheels have drum brakes. For them, both reservoirs are of the same size. The same is true for the only American-built cars with four-wheel disc brakes: the Corvette since 1965, a few Camaros and, since 1975, the Imperial.

When all four wheels use the same kind of brakes, even though the system is dual for reasons of safety, no differential in hydraulic pressure between the front and rear systems is required or even desired. But, for systems with rear-wheel drum brakes and front-wheel disc brakes, not only are greatly different pressures required but the time of pressure application is important. Thus, we have need for differentiating valves in the hydraulic system as well as the variations in reservoir size. These valves are discussed later.

Figure 2-7 shows the working parts of a typical dual-system master cylinder, which is essentially two master cylinders combined into one housing. On the cylinder shown, which happens to be for a system with rear-wheel drum brakes and front-wheel disc brakes, the various parts of the assembly are pointed out.

The cover and the rubber diaphragm are intended to provide an airtight seal for the reservoirs—airtight to keep dust, moisture, and all extraneous materials from the atmosphere out of the fluid. The cover may be held in place by a bail or bolt, and must be vented to allow atmospheric pressure to act upon the diaphragm.

BRAKE FLUID RESERVOIR

The reservoirs—*primary* for the front wheels and *secondary* for the rear wheels—are for one purpose: to hold in reserve the hydraulic fluid needed, in addition to that already in the system, for braking action. If there were no surplus fluid in the reservoir, air in the system would be a constant problem. And the very thing that makes any hydraulic system effective is the fact the liquid medium is almost incompressible, while air can be compressed a great deal. So, if enough air enters a hydraulic system, as it sometimes does, the brake pedal can be pushed to the floor and all that happens is that this air is compressed. We'll discuss this problem and what to do about it elsewhere in this book.

Two ports in each reservoir allow fluid to flow into and out of the cylinder base from the reservoir. One of these ports is called a *compensating* port; the other doesn't seem to have a name, but we sometimes call it the "lubricating" hole, since its main function seems to be to allow brake fluid to lubricate the cylinder and piston at all times.

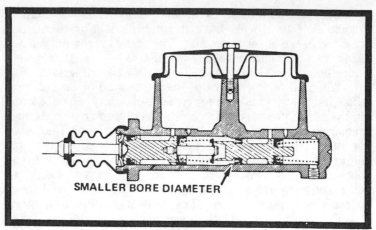

SMALLER BORE DIAMETER

Fig. 2-8. Typical dual-system master cylinder with "step-bore" cylinder.

Speaking of pistons, most assemblies are much like that shown in Fig. 2-7, with both the primary and the secondary cylinders of the same bore and the two pistons of the same caliber. There are in use, however, a few "step-bore" dual-system master cylinders. These cylinders are essentially the same as all others, except that the forward, or secondary, part of the bore is somewhat smaller than the primary part. (See Fig. 2-8.)

Otherwise, not even those hybrid master cylinders that accommodate rear-wheel drum brakes and front-wheel disc brakes (which we refer to as "disc-brake master cylinders") differ much from their four-wheel drum brake counterparts.

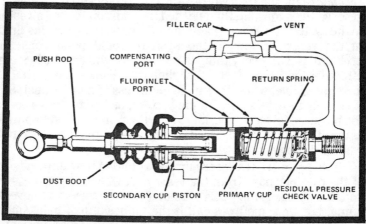

Fig. 2-9. Typical single-system master cylinder with residual pressure check valve.

RESIDUAL-PRESSURE CHECK VALVE

The outlet from the disc-brake reservoir never contains a residual-pressure check valve (Fig. 2-9); however, this is not an infallible clue, since some drum-brake reservoirs also lack this valve.

The purpose of this valve is to retain a slight pressure—about 5 to 25 psi—at all times. This constant pressure tends to keep air out of the system, and it maintains a certain amount of tension against the return springs, which permits a more ready response when pressure is reapplied for braking action. But disc brakes employ no return springs and therefore require no residual-pressure check valve.

And, indeed, in recent models some car builders have done away with residual check valves in brake systems altogether. Of course, the big swing to disc brakes has been a factor in this change, but more importantly, *cup expanders* (Fig. 2-10), have provided a better (or less expensive) means of excluding air from the system. These expanders work to prevent air from sneaking past the lips of cups in wheel cylinders and into the system. Supposedly, their use greatly reduces cup fatigue.

VACUUM BOOSTER (POWER BRAKES)

If your car has *power brakes* it employs an air or vacuum booster. (The possible exception to this statement—the Bendix

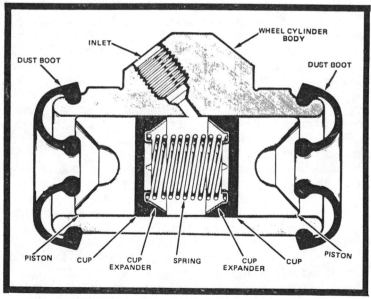

Fig. 2-10. Typical cup expander installation.

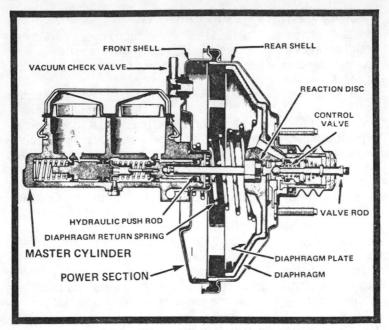

Fig. 2-11. Typical Bendix power brake "Master-Vac."

"Hydro-Boost" power brake—is covered briefly at the conclusion of this discussion.) Its function, quite naturally, is to give you a great deal more braking with a less strenuous effort. Those of you whose experience includes wrestling a horse-drawn, hand-braked wagon down a mountain road can remember how it took every ounce of strength you had and all the help the horses could give you to keep that wagon from running over them and away with you. And no one has to tell you that brakes have come a long way since those "good old days." But, you don't have to be as old as Methuselah to have driven a car without power brakes, for there are many around today; and indeed many, if not most, smaller American-built cars leave the factory today without this feature.

Because power brakes are one more plus for good braking, we tend to favor their use whenever it is practical. But, there we have the key word, *practical*. And, because the power for power brakes is derived from the partial vacuum now demanded for operation of a whole fistful of accessories and emission-control devices—we sometimes find there just isn't enough vacuum to go around. It is a combination of the partial vacuum created within the engine's intake manifold and the rush of atmospheric pressure to fill this partial vacuum that provides the power to operate power-brake vacuum boosters.

Power brakes came into fairly common use in the mid-1950s. And, prior to 1964, some manufacturers used what was called an "atmospheric-suspended" booster. In general, this type worked somewhat opposite to the more common "vacuum-suspended" type employed by all American manufacturers since 1964. So, because the odds must be at least 99 to 1 that if your car has power brakes it has the latter type, we have elected to limit this discussion to the vacuum-suspended system.

Power-brake vacuum boosters are all pretty much alike. In fact, there are only three major manufacturers: Bendix, Delco-Moraine, and Midland-Ross. And, as with car brake systems in general, the Bendix system is by far the most popular—their *Master-Vac* unit has been widely used since 1958.

Perhaps the most significant difference in vacuum boosters is in the manner in which different ones are mounted on the firewall (that metal wall which separates the engine compartment of your car from the passenger compartment).

The Bendix Master-Vac (Fig. 2-11) consists of two main components: (1) the master cylinder and (2) the power section.

The power section consists of two metal shells, called the front shell and rear shell, that form a single vacuum chamber;

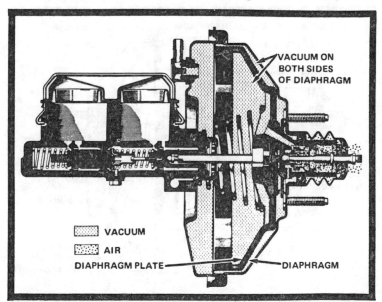

Fig. 2-12. Single-diaphragm "Master-Vac" in cross section, with brake pedal at rest.

however, a diaphragm and a diaphragm plate separate this chamber into two parts. Also, the power section utilizes a vacuum check valve, as well as a hydraulic pushrod, a diaphragm-return spring, a reaction disc, a valve rod, and a control valve. And, finally, this entire assembly is attached to the intake manifold by a vacuum hose.

When the brake pedal is at rest, equal engine vacuum is present on both sides of the diaphragm (Fig. 2-12) and the diaphragm is also at rest. As the brake pedal is depressed the control valve closes off the supply of vacuum to the rear of the diaphragm. Simultaneously, the valve vents atmospheric pressure into this rear section of the chamber (Fig. 2-13). Thus, with the partial vacuum in front of the diaphragm and atmospheric pressure behind it, both the diaphragm and plate move forward with a force directly proportional to the difference in atmospheric pressure. And this force can be—if the pressure difference is significant—virtually irresistible. As the plate moves forward, the pushrod attached to it also moves forward, forcing the piston into the master-cylinder bore. As we've already discovered, this movement of the piston builds up the hydraulic pressure that applies the wheel brakes.

The single-diaphragm Master-Vac is shown in Fig. 2-14. This unit provides the same braking power per square inch of

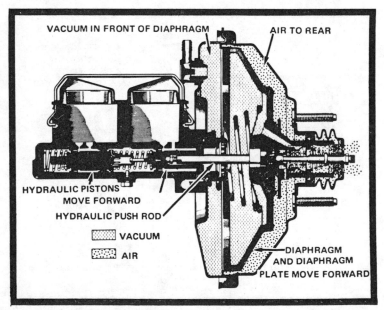

Fig. 2-13. Single-diaphragm "Master-Vac" in cross section with brake pedal depressed.

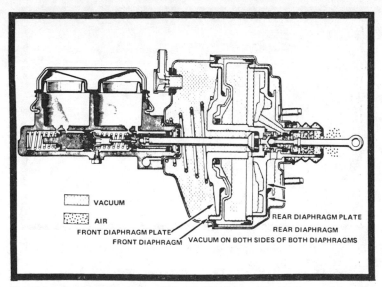

VACUUM
AIR
FRONT DIAPHRAGM PLATE
FRONT DIAPHRAGM
REAR DIAPHRAGM PLATE
REAR DIAPHRAGM
VACUUM ON BOTH SIDES OF BOTH DIAPHRAGMS

Fig. 2-14. Tandem-diaphragm "Master-Vac" in cross section, with brake pedal at rest.

diaphragm, but occupies considerably less mounting space. Cross sections of this unit are shown in Figs. 2-14 and 2-15, and close scrutiny of these figures should reveal that it operates in essentially the same manner as the single-diaphragm unit, except that the two diaphragms work together to provide force

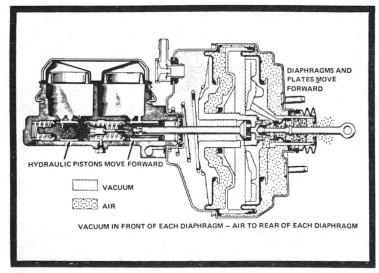

DIAPHRAGMS AND PLATES MOVE FORWARD
HYDRAULIC PISTONS MOVE FORWARD
VACUUM
AIR
VACUUM IN FRONT OF EACH DIAPHRAGM — AIR TO REAR OF EACH DIAPHRAGM

Fig. 2-15. Tandem-diaphragm "Master-Vac" in cross section, with brake pedal depressed.

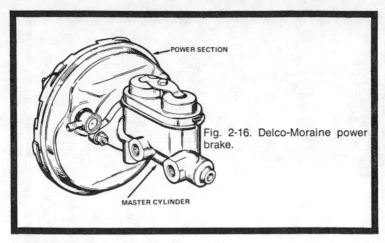

POWER SECTION

Fig. 2-16. Delco-Moraine power brake.

MASTER CYLINDER

to the pushrod—that is, the two plates are connected by a hub and move as a unit.

Delco-Moraine (Fig. 2-16) and Midland-Ross (Fig. 2-17) units are constructed and operate essentially in the same manner as the Bendix unit.

Further on in this book we will provide you with a simple troubleshooting chart that will allow you to discover what, if anything, might be wrong with your vacuum booster. But we can urge you now never to attempt to overhaul the power section of a power brake unless you have all the tools and the special test bench required. Even then, you usually can replace one at less than the cost of repair.

Most power brakes on automobiles today use engine vacuum to operate a booster or power section. Bendix's new *Hydro-Boost* departs from that principle and uses hydraulic pressure developed by the power-steering pump to operate a booster that helps to apply the brakes.

The addition of emission-control devices and vacuum-operated accessories in recent years has reduced the

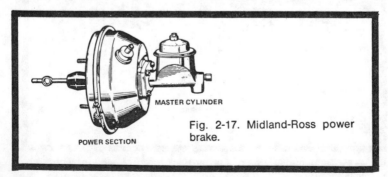

MASTER CYLINDER

Fig. 2-17. Midland-Ross power brake.

POWER SECTION

engine vacuum available to operate a vacuum power brake. As a result, power brakes have grown larger to provide the needed boost and more difficult to fit into the limited space available under the hood. Hydro-Boost is smaller, more compact. And it provides the boost necessary to produce the relatively high pressure required in hydraulic brake systems.

The other brake components used with Hydro-Boost, from the master cylinder to the calipers and wheel cylinders, are the same as those used with vacuum-power brakes. And operation of the Hydro-Boost is quite simple. Hydraulic fluid from the power-steering pump flows through the Hydro-Boost and then to the power-steering gear.

When the brake pedal is depressed, a portion of the fluid is diverted into the power section of the Hydro-Boost. The fluid moves a piston that operates the master cylinder and applies the brakes. Sufficient fluid for power steering continues to flow through the unit.

A hydraulic accumulator that stores fluid under pressure provides a safety feature. The pressurized fluid is sufficient for two or three power-assisted brake applications in the event of a power-steering pump failure.

And even after the pressurized fluid is gone, the brakes can be applied manually by pressing on the brake pedal with greater force.

Valve and Switch Fundamentals

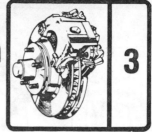

3

A number of valves and switches may be included in your car's brake system, in addition to the residual pressure check valve discussed in the previous section. Let's look at a few of them.

STOPLIGHT SWITCH

When you apply your brakes, the red stoplights at the rear of your car should light up. In most states these lights must be red and noticeably brighter than your running lights, and they must work. More often than not, if they aren't working properly, the problem is a bulb or fuse, but occasionally your stoplight switch may go bad.

This spring-loaded electrical switch comes in two general types: (1) mechanically operated switches, used on most late model cars, and (2) hydraulically operated switches, used on a few older model cars.

The mechanical switches, shown in Fig. 3-1, are operated by contact with some part of the brake-pedal lever or a bracket attached to it. Depressing the "button" on one of these switches simply closes the electrical circuit to the stoplights. Releasing the brake pedal releases the button and a spring pushes it up, breaking the electrical circuit and turning the stoplights off again. Although the principle is the same, some cars—such as late-sixties GM products—use the opposite arrangement: In the normal (unbraked) condition, the brake-pedal shaft holds a switch in a spring-loaded *engaged* position, which prevents voltage from being supplied to the stoplights. When the brake pedal is pushed the switch is released, allowing voltage to be delivered to the brake lights.

The hydraulic switches, shown in Fig. 3-2, work in essentially the same manner. One of these switches can be mounted so that increased hydraulic pressure in the brake system, caused by applying foot pressure to the brake pedal, presses on its diaphragm which, in turn, forces an electrical contact, completing the circuit to the stoplights. When

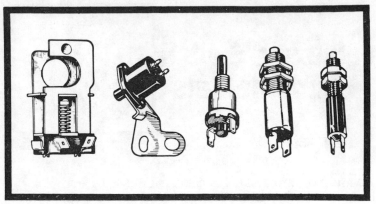

Fig. 3-1. Mechanically operated stoplight switches.

hydraulic pressure is released, a spring forces the contact open in the switch and turns off the stoplights.

Stoplight switches may wear, break, and corrode. More often, they'll shift in position so that the at-rest brake pedal does not contact the switch properly. Sometimes a hydraulic switch can develop a leak. How to discover what problem you may have with your switch and what to do about it are discussed under the appropriate headings in Chapter 6, as is the brake-warning light switch.

BRAKE-WARNING LIGHT SWITCH

Usually, when we mention this switch by name the nonmechanic car owner believes we are referring to the light that warns him or her that the parking brake is *on*—a light that is actuated by a device much like the mechanical stoplight switch mentioned above. But the brake-warning light switch serves a much more serious function: it warns you if either section of your dual system fails. For example, should your brake system develop a debilitating hydraulic leak at either rear wheel cylinder, causing pressure in the secondary (rear) section to be substantially less than that in the primary section, the brake-warning light switch would be activated and the brake-warning light on your instrument panel would come on.

Brake-warning light switches first came into common use in 1967, along with common adoption of the mandatory dual system and dual-system master cylinder. And, while all such switches serve identical purposes, three separate designs have experienced a certain amount of popular use. Perhaps the simplest is the GM type shown in Fig. 3-3, which is identifiable by its single piston and two centering springs.

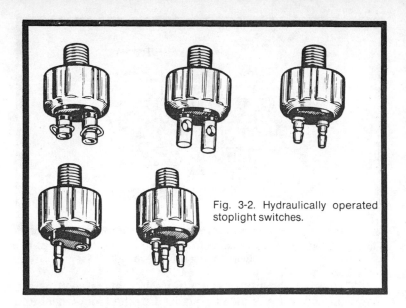

Fig. 3-2. Hydraulically operated stoplight switches.

Single-Piston Switch With Centering Springs

In this design, the hydraulic lines are attached as indicated in Fig. 3-4. And, when all is as it should be, hydraulic pressure is the same at both ends of the piston and the piston remains centered, held there by the two centering springs. As long as the piston is not in contact with the switch terminal, the brake-warning light does not come on. But, should there be a failure—a severe loss of hydraulic pressure—in the front-wheel system, pressure from the rear-wheel system forces the piston to the left as shown in Fig. 3-5. As the shoulder of the piston makes contact with the terminal, the warning light comes on. Had the rear-wheel system failed, the piston would have moved in the opposite direction.

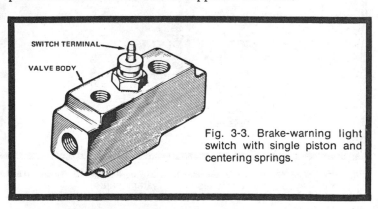

SWITCH TERMINAL

VALVE BODY

Fig. 3-3. Brake-warning light switch with single piston and centering springs.

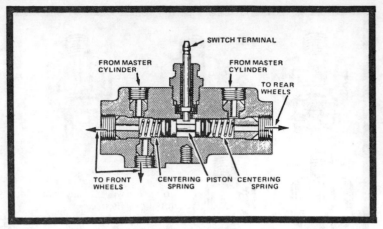

Fig. 3-4. Cross section of brake-warning light switch with single piston and centering springs: system in normal operation.

The simple beauty of this design is that as soon as the hydraulic pressure is again equalized in the front-wheel and rear-wheel systems, the centering springs force the piston to center once more, and the warning light goes out. This feature is not found in the more complex two-piston switch.

Two-Piston Switch With Centering Springs

This switch (Fig. 3-6), combines some of the best and some of the worst features of the other two popular switch types discussed. It operates much like the one discussed above, except that it has a moveable switch terminal that makes electrical contact by dropping into its housing as the cylinders are forced off center (Fig. 3-6 and 3-7) by a variation in the

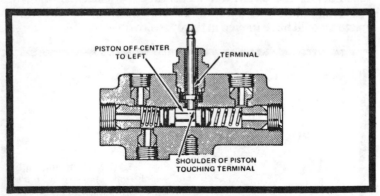

Fig. 3-5. Cross section of brake-warning light switch with single piston and centering springs: front-wheel brake system failure.

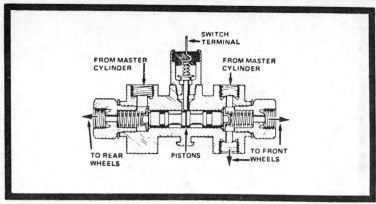

Fig. 3-6. Cross section of brake-warning light switch with two pistons and centering springs: system in normal operation.

pressure between the front-wheel and rear-wheel hydraulic systems. But, in this switch, the pistons cannot be recentered by the centering springs until the switch terminal is manually returned to its original position.

Generally this switch is found only on cars and light trucks manufactured by American Motors, International, and Kaiser-Jeep. The terminal should be removed from it prior to bleeding any parts of the system, or else it is apt to be damaged.

Single-Piston Switch Without Centering Springs

Except for its lack of centering springs, this type of switch has essentially the same parts as the first switch we discussed; however, the general design of the switch terminal is also different. What happens in the event of a system failure is shown in Figs. 3-8 and 3-9. In Fig. 3-9 the terminal has been

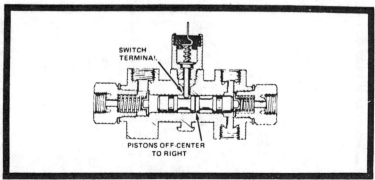

Fig. 3-7. Cross section of brake-warning light switch with two pistons and centering springs: rear-wheel brake system failure.

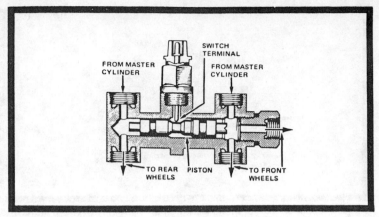

Fig. 3-8. Cross section of brake-warning light switch with single piston and no centering springs: system in normal operation.

forced upward by the piston shoulder and an electrical circuit has been closed in the terminal housing to turn on the brake-warning light.

But, as with the two-piston switch, the warning light will remain on until you center the switch. Unlike the two-piston switch, however, this one must be recentered by bleeding the hydraulic system opposite the one that failed.

Since 1970, many cars with front-wheel disc brakes have used a combination valve/switch. On some the brake-warning light switch is a single assembly with the proportioning valve, and on others it is combined with the metering valve or both.

METERING VALVE

The function of a metering valve is to help equalize braking pressure between the front-wheel system and the

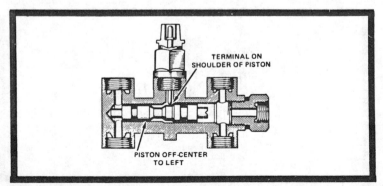

Fig. 3-9. Cross section of brake-warning light switch with single piston and no centering springs: rear-wheel brake failure.

rear-wheel system, especially in light braking actions. It does so by preventing hydraulic pressure from being applied to the front disc brakes until sufficient pressure has been applied to the rear-wheel brakes to overcome the shoe-return springs. Theoretically, if the total system builds to at least 75 psi (but no more than 135 psi), then both the rear-wheel shoes and the front-wheel pads will take hold at the same time.

Metering valves are nonrepairable and non-adjustable—just replaceable. And they come in two general types: (1) the *push* type shown in Fig. 3-10 and (2) the *pull* type shown in Fig. 3-11. The purpose of their valve stems, whether *push* or *pull*, is to allow a bypass of fluid while the system is being bled. However, if your brakes are bled manually, pressure developed by depressing the brake pedal is adequate to overcome the metering valve; it may be ignored, as may the proportioning valve.

PROPORTIONING VALVE

This valve, situated in the rear-wheel system, serves to send the lion's share of hydraulic pressure to the front-wheel disc brakes during heavy braking actions. Otherwise, the rear wheels might be locked in a skid before adequate pressure arrived at the front-wheel disc brakes. Thus, the proportioning valve (Fig. 3-12), also serves to maintain a proper braking balance between the two systems. And it too is a nonadjustable, nonrepairable valve.

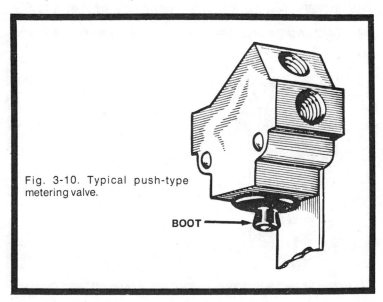

Fig. 3-10. Typical push-type metering valve.

BOOT →

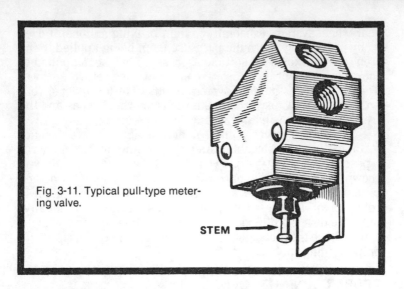

Fig. 3-11. Typical pull-type metering valve.

STEM →

COMBINATION VALVE

Beginning with the 1970 models, many cars have used combination valves—that is, one valve assembly that performs two or three of the functions described above.

Two-Function Valve

There are a couple of variations of the two-function valve in use. One combines the brake-warning light switch and the proportioning valve into a single assembly, as shown in Fig.

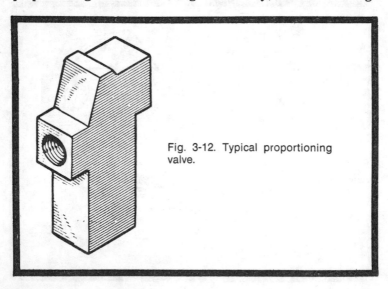

Fig. 3-12. Typical proportioning valve.

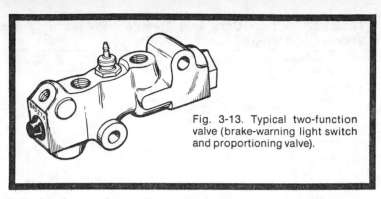

Fig. 3-13. Typical two-function valve (brake-warning light switch and proportioning valve).

3-13. Another packages the brake-warning light switch with the metering valve, as shown in Fig. 3-14. It should be noted that this latter combination may be quite similar in appearance to some three-function valves—the difference being in their internal makeups.

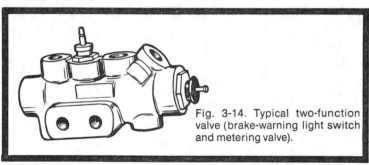

Fig. 3-14. Typical two-function valve (brake-warning light switch and metering valve).

Three-Function Valve

This valve simply combines all the three functions mentioned above (brake-warning light switch, metering valve,

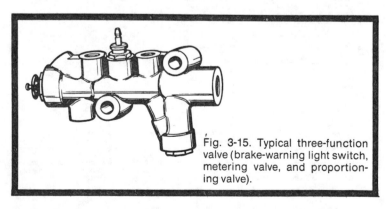

Fig. 3-15. Typical three-function valve (brake-warning light switch, metering valve, and proportioning valve).

Table 3-1. Which Cars Use Residual-Pressure Check Valves? And Where?

Make and Model	Year	Metering	Proportioning	Three-Function	Two-Function (1)	Two-Function (2)
AMERICAN MOTORS						
All	1965-68	–	–	–	–	–
Ambassador & Rebel, 8-Cyl.	1969-70	–	Yes	–	–	–
Ambassador & Rebel, 6-Cyl.	1969-70	–	–	–	–	–
AMX, Javelin & Hornet	1969-70	–	Yes	–	–	–
All	1971	–	–	Yes(B)	–	–
CHECKER MOTORS						
All	1969-70	Yes(A)	Yes	–	–	–
All	1971	–	–	Yes(A)	–	–
CHRYSLER CORPORATION						
Barracuda, Dart & Valiant	1965-69	–	Yes	–	–	–
Barracuda	1970	Yes(B)	Yes	–	–	–
Dart & Valiant	1970	–	Yes	–	–	–
Barracuda	1971	Yes(B)	–	–	–	Yes
Dart & Valiant	1971	–	–	–	–	Yes
Charger, Coronet, Satellite & Belvedere	1966-69	–	Yes	–	–	–
Road Runner & Super Bee	1969	–	Yes	–	–	–
Charger, Coronet, Challenger, Satellite, Belvedere, Road Runner & Super Bee	1970	Yes(B)	Yes	–	–	–
Charger, Coronet, Challenger, Satellite, Road Runner & Super Bee	1971	Yes(B)	–	–	–	Yes
Fury, Polara, Monaco & Chrysler	1966-69	Yes(A)	–	–	–	–
Imperial	1967-69	Yes(A)	–	–	–	Yes
Fury, Polara, Monaco, Chrysler & Imperial	1970	Yes(B)	–	–	–	–
Fury, Polara, Monaco, Chrysler & Imperial	1971	–	–	–	Yes(B)	–
FORD MOTOR COMPANY						
Mustang, Fairlane & Falcon	1965-69	–	Yes	–	–	–
Ranchero	1967-69	–	Yes	–	–	–
Torino	1968-69	–	Yes	–	–	–
Falcon	1970	–	–	–	–	Yes
Mustang, Torino & Ranchero	1970-71	–	–	–	–	Yes
Ford, Mercury & Thunderbird	1965-69	–	Yes	–	–	–
Mark III	1968-69	Yes(B)	Yes	–	–	–
Lincoln Continental	1965	Yes(A)	Yes	–	–	–
Lincoln Continental	1966-69	Yes(B)	Yes	–	–	–
Ford, Mercury, Thunderbird, Mark III & Lincoln Continental	1970-71	Yes(B)	–	–	–	Yes
Pinto	1971	–	–	–	–	–
F-250 & F-350 trucks	1968-71	Yes(A)	–	–	–	–
GENERAL MOTORS CORPORATION						
Buick, Riviera, Special & Skylark	1967-70	Yes(A)	–	–	–	–
Buick, Riviera, Special & Skylark	1971	–	–	Yes(A)	–	–
Buick & Special station wagons	1971	–	–	–	Yes(A)	–
Cadillac	1968-70	Yes(A)	–	–	–	–
Cadillac Eldorado	1967-70	Yes(A)	Yes	–	–	–
Cadillac & Eldorado	1971	–	–	Yes(A)	–	–
Cadillac Commercial & Limousine	1971	–	–	–	Yes(A)	–
Chevrolet	1967-70	Yes(A)	–	–	–	–
Chevrolet	1971	–	–	–	Yes(A)	–
Camaro	1967-70	Yes(A)	Yes	–	–	–
Chevelle, Chevy II & Nova	1967-69	Yes(A)	–	–	–	–
Chevelle, Chevy II & Nova, Monte Carlo & El Camino	1970	Yes(A)	Yes	–	–	–
Camaro, Chevelle, Nova, Monte Carlo & El Camino	1971	–	–	Yes(A)	–	–
Corvette	1965-71	–(4)	–	–	–	–
Vega	1971	–	–	–	–	–
Chevrolet Truck 10, 20 & 30 Series	1971	–	–	Yes(A)	–	–
Oldsmobile & Toronado	1967-70	Yes(A)	Yes	–	–	–
Oldsmobile F-85	1967-70	Yes(A)	–	–	–	–
Oldsmobile, Toronado & F-85	1971	–	–	Yes(A)	–	–
Oldsmobile station wagon	1971	–	–	–	Yes(A)	–
Pontiac	1967-70	Yes(A)	–	–	–	–
Tempest & Grand Prix	1967-69	Yes(A)	–	–	–	–
Tempest & Grand Prix	1970	Yes(A)	Yes	–	–	–
Firebird	1967-68	Yes(A)	Yes(3)	–	–	–
Firebird	1969-70	Yes(A)	Yes	–	–	–
Pontiac	1971	–	–	–	Yes(A)	–
Tempest, Grand Prix & Firebird	1971	–	–	Yes(A)	–	–

(1) — Combination metering valve and brake-warning light switch.

(2) — Combination proportioning valve and brake-warning light switch.

(3) — With air conditioning.

(4) — 1967-68 models with heavy-duty disc brakes have metering valve in line to rear brakes.

(A) — Push type

(B) — Pull type

and proportioning valve) into one assembly such as that shown in Fig. 3-15.

It should be noted that combination valves, like the single-function valves mentioned previously, are nonadjustable and nonrepairable. If yours is defective, the only recourse is to replace it.

WHICH CARS USE WHICH VALVES?

Table 3-1, current only through 1971, shows which cars use which valves and in what combinations. As this table suggests, valve use had become fairly standardized by 1971, and your later-model car is apt to be equipped similarly; however, a quick check with your parts salesman can confirm the valve arrangement in your car.

Drum Brake Fundamentals

4

The history of drum brakes on American cars dates almost from the first car. The popular *duoservo* brake came into use on the Chevrolet some fifty years ago. In that same era Chrysler introduced the hydraulic brake, which almost overnight, made mechanical brakes obsolete.

Today the drum brake is overshadowed by the disc brake. The disc is superior in almost every way (except in terms of part costs!) and is, no doubt, the brake configuration of the future. It is fade-resistant, impervious to water, and functionally simple. However, drum brakes continue to be used, particularly on the rear wheels.

Why?

There are a couple of practical reasons, both of which are being slowly overcome.

First, installation of rear-wheel disc brakes creates a new problem: what to do about the parking brake? As we will soon see, the parking brake on most American cars is nothing more than a simple mechanical linkage by which the rear-wheel drum brakes are actuated—a simple, effective, and inexpensive arrangement. But, with rear-wheel disc brakes, a completely separate parking brake system—such as the Corvette system shown in Fig. 4-1 or the old Chrysler system shown in Fig. 4-2—becomes necessary. And such systems are cumbersome and expensive.

A second reason why disc brakes have evolved more slowly in America than in Europe is that American driving habits are quite different from the driving habits of Europeans. Americans tend to put a lot of miles on cars and except in a few cases they expect to put a lot of miles on a car before that first brake job becomes necessary. Americans have one other peculiarity that only recent generations of drivers are beginning to overcome: *riding the brake pedal!* The habit is not too damaging to spring-loaded drum brakes but it's extremely detrimental to disc brakes. About a decade ago, I purchased a new European-built Volvo with four-wheel disc brakes. With only 5500 miles on the speedometer, I had to

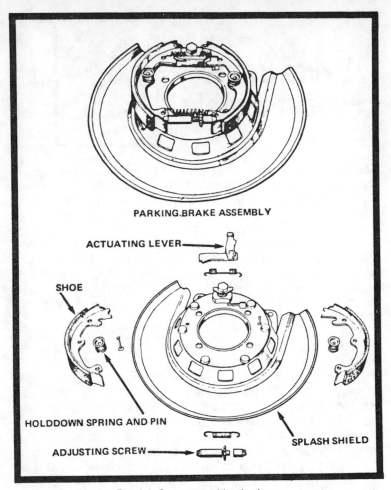

PARKING BRAKE ASSEMBLY

ACTUATING LEVER

SHOE

HOLDDOWN SPRING AND PIN

ADJUSTING SCREW

SPLASH SHIELD

Fig. 4-1. Corvette parking brakes.

have all the brake pads replaced—I'd worn them out riding the brake pedal. Disc brakes have no return springs to offset light foot pressure. Even so, I complained to the dealer from whom I purchased the Volvo, but he stuck by his service charge, explaining that many American purchasers wore out their first set of pads rapidly. "That's the way it is," he concluded.

My experience with the disc brakes on that Volvo reflected more on the driver than on the car, as experience with other disc brakes have since proven. And it should be pointed out that American-built cars with the current popular arrangement of rear-wheel duoservo drum brakes and front-wheel disc brakes should not need servicing for at least

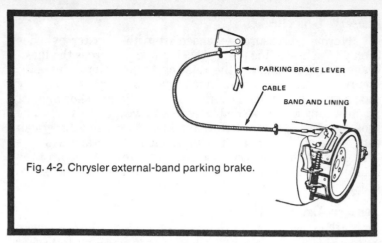

Fig. 4-2. Chrysler external-band parking brake.

40,000 miles. Twice that mileage is more common, and three times is not uncommon.

THE PARTS OF YOUR DRUM BRAKES

At first glance a drum brake seems to be a rather complex assembly of parts and pieces held in place by a bunch of springs. And, indeed, a considerable history of engineering and experimentation has gone into making this brake system what it is today—a fairly uncomplicated, easy-to-service, but highly efficient means of stopping your car and keeping it that way, with little more than a light touch of your toe.

What, then, are the main parts of a drum-brake assembly? Most consist of the following:

- A backing or *support* plate.
- Two brake shoes, attached to the backing plate.
- Two brake shoe linings (the primary shoe lining is usually shorter than the secondary shoe lining).
- Return springs and shoe holddown hardware.
- An automatic shoe adjuster, to compensate for shoe lining wear (automatic brake adjusters, first introduced on 1957 models, have been used on all American-built cars since 1963).
- A drum.
- For rear wheels, a parking-brake strut.

SERVO DRUM BRAKES

Drum brakes have evolved through numerous designs, but all such brakes used on modern American-built cars are of two types: *duoservo* and *nonservo*.

Duoservo Drum Brakes

Hydraulic pressure—extended from the master cylinder through the brake line to the wheel cylinder—forces the lined brake shoes out against the rotating brake drum (Fig. 4-3). The rotating motion of the drum serves to wedge the primary shoe against it and the movement of the primary shoe against the adjusting screw assembly serves to wedge the secondary shoe against the drum. Before the late 1950s a few American cars were still being built with single-servo brake systems (that is, only the primary shoe became "wedged" against the drum). And even today a couple of small cars use nonservo brakes.

Nonservo Drum Brakes

As late as 1975, the Chevrolet *Vega* and the American Motors *Gremlin* were using nonservo rear-wheel drum brakes (Fig. 4-4). These brakes lack the "wedging" force discussed above, but function adequately on cars. In fact, the front-wheel disc brakes used by both Vega and Gremlin provide adequate stopping power for these cars, and the rear-wheel nonservo drum brakes are there as much to provide smoothness and straightness to breaking as to enhance overall stopping power. And, of course, the nonservo brake assemblies are somewhat less expensive than their counterparts.

BRAKES IN OPERATION

When you apply the brakes the brake shoes are forced outward against the rotating drum. On servo brakes, as the

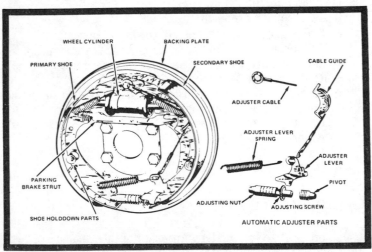

Fig. 4-3. Typical duoservo drum brake (with drum removed).

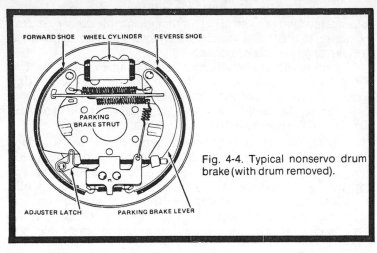

Fig. 4-4. Typical nonservo drum brake (with drum removed).

lining contacts the drum, the primary shoe tends to move in the direction of wheel rotation (Fig. 4-5). As it moves, ever so slightly with the drum, force from the primary shoe is transmitted through the adjusting screw to the secondary shoe whose lining has also been extended into contact with the drum, and is now forced into tightened contact. Thus, by means of this self-serving process, brake efficiency is greatly increased.

An opposite action occurs in reverse braking with the secondary shoe wedging the drum. Reaction through the adjusting screw forces the primary shoe into tightened contact.

Before installation of automatic adjusters in drum brakes, keeping brakes adjusted was a continual hassle, especially on cars in hard use such as mountain driving. But for the most

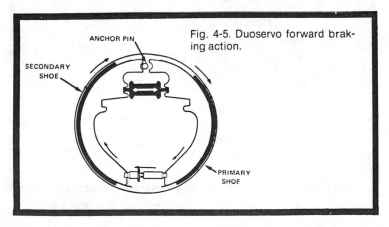

Fig. 4-5. Duoservo forward braking action.

part, manual brake adjustment is a thing of the past, for the automatic adjuster is an efficient though simple device. It uses the action of the brake shoes during reverse brake application to make any needed adjustment.

A typical automatic adjuster is shown in Fig. 4-6. The eye of the adjuster cable fits over the anchor pin on the backing plate. And the hook on the other end engages the adjuster lever. Then the cable loops over the cable guide which is attached to the secondary shoe.

During each reverse brake application, as the secondary shoe tries to rotate with the drum, the attached cable guide moves away from the anchor pin and pulls upward on the adjuster lever. Upon release, if the lining has worn sufficiently to allow the lever to engage the next tooth on the adjusting-screw star wheel, the adjuster-lever spring pulls the lever back to its normal position, turning the star wheel one notch.

No adjustment occurs, however, if the adjuster lever does not move far enough to engage the next tooth on the star wheel.

It should be noted that automatic adjusters on different manufacturers' brakes are not identical. As can be seen in Figs. 4-6 through 4-12, however, they are similar.

WHEEL CYLINDERS

Wheel cylinders work in a manner opposite that of the master cylinder. Master cylinders receive energy transmitted from your foot mechanically (that is, by mechanical linkage) through the pedal and clevis to the master cylinder piston. And

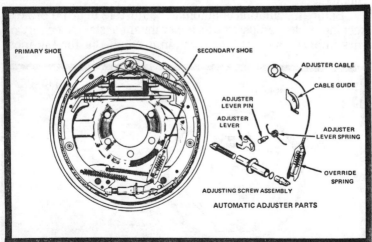

Fig. 4-6. Typical duoservo brake with Bendix automatic adjuster (left rear).

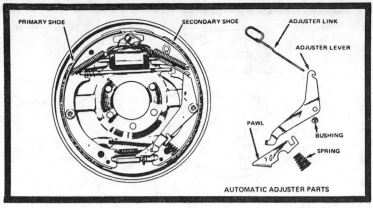

Fig. 4-7. Bendix BX brake with automatic adjuster (left rear).

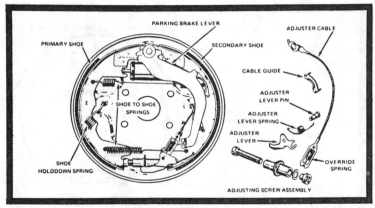

Fig. 4-8. Delco-Moraine brake with automatic adjuster (left rear) with lever and pawl.

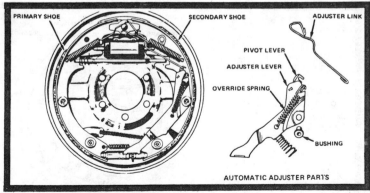

Fig. 4-9. Delco-Moraine brake with automatic adjuster with pivot lever and override springs (left rear).

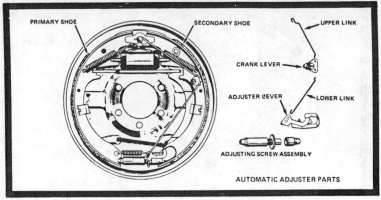

Fig. 4-10. Wagner brake with automatic adjuster (left rear).

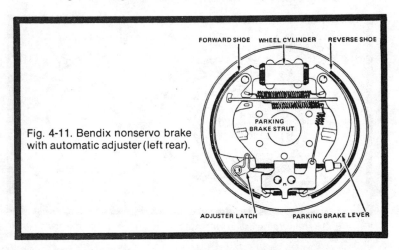

Fig. 4-11. Bendix nonservo brake with automatic adjuster (left rear).

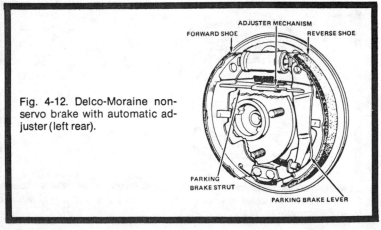

Fig. 4-12. Delco-Moraine nonservo brake with automatic adjuster (left rear).

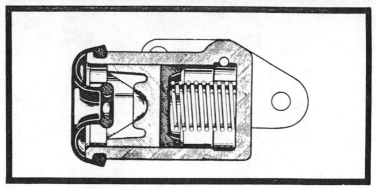

Fig. 4-13. Single-piston wheel cylinder.

the piston applies this energy to the hydraulic fluid in the master cylinder. The hydraulic fluid transmits energy (via the column of hydraulic fluid in the line) from the master cylinder to wheel cylinders. Then, in the wheel cylinder, this energy once more is transferred from the hydraulic fluid to the wheel cylinder piston—and once more it becomes mechanically transmitted energy as the cylinders force the lined shoes outward against the drums.

Wheel cylinders are second only to shoe linings as a major source of brake problems. Three basic types of cylinders have been used with drum brakes:

- Single piston.
- Double piston, step bore.
- Double piston, straight bore.

Single-Piston Cylinders

The single-piston wheel cylinder, used by some American-built cars as recently as 1957, has only one cup, one piston, and one dust boot, as shown in Fig. 4-13.

Double-Piston, Step-Bore Cylinder

The double-piston, step-bore wheel cylinder differs from the more common straight-bore type in that the cylinder is bored in two different dimensions, as shown in Fig. 4-14. To accommodate this feature, two cups, pistons, and dust boots of different sizes are required.

Double-Piston, Straight-Bore Cylinders

This type of wheel cylinder has been used on all American-built cars since 1967. The cross section (Fig. 4-15)

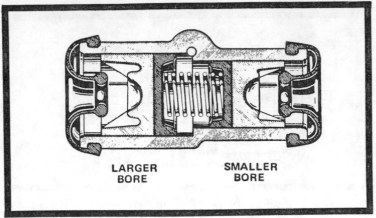

LARGER
BORE

SMALLER
BORE

Fig. 4-14. Double-piston, step-bore wheel cylinder.

shows all the typical features of a modern wheel cylinder except the bleeder screw, of which several types are illustrated in Fig. 4-16.

Figures 4-17 and 4-18 illustrate two other types of wheel cylinders in which the method of attaching the dust boots to the cylinder body varies slightly.

How Wheel Cylinder Works

The cavity between the cups (Fig. 4-15) must remain full of hydraulic fluid at all times. Then, when you depress the brake pedal, more fluid is forced into this cavity. The cups are forced outward against the pistons, which move the shoe links and brake shoes outward, forcing the shoe linings into contact with the drum. Thus, brake action occurs.

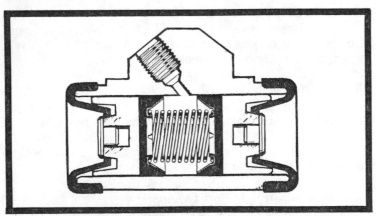

Fig. 4-15. Double-piston, straight-bore wheel cylinder.

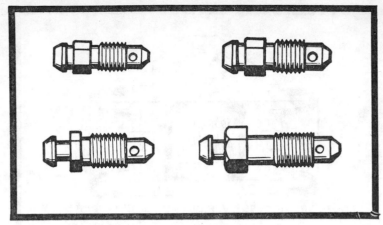

Fig. 4-16. Several typical bleeder screws.

When foot pressure is released, return springs pull the shoes back to their original positions, simultaneously returning the pistons and cups to their original positions, forcing the excess hydraulic fluid out of the wheel-cylinder fluid cavity, through the line, and back into the master cylinder.

As shown in Fig. 4-19, some shoe webs rest directly upon the wheel cylinder pistons, eliminating the shoe links in the more common arrangement illustrated in Fig. 4-15.

BRAKE SHOES AND LININGS

The major source of brake failure is nothing more complicated than worn shoe linings. Thus, linings and the shoes that hold them are the single most important maintenance or service item in any drum-brake system. A

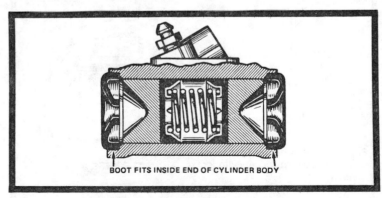

Fig. 4-17. Wheel cylinder with internal boot.

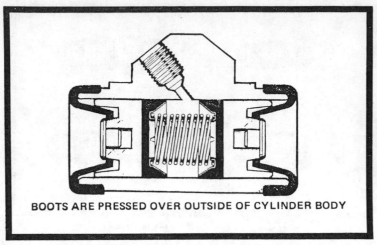

BOOTS ARE PRESSED OVER OUTSIDE OF CYLINDER BODY

Fig. 4-18. Wheel cylinder with external press-on boot.

brief investigation of these items, their individual construction and their collective function is in order.

Brake Shoe Construction

The brake shoe must support the lining and force it against the drum surface when brake pressure is applied—evenly translating pressure from the system into braking power. A typical brake shoe is shown in Fig. 4-20. And, though shoes for a Ford may not be identical to those used on a Chevrolet, they are similar in appearance. All brake shoes have many points in common.

The shoe rim (the mounting surface for the lining) is welded to the web to provide a rigid or slightly flexible

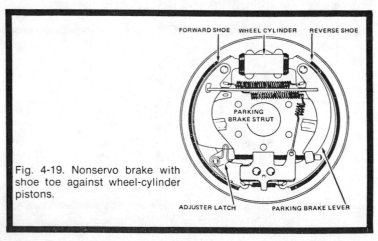

Fig. 4-19. Nonservo brake with shoe toe against wheel-cylinder pistons.

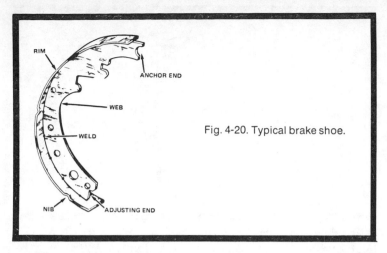

Fig. 4-20. Typical brake shoe.

platform for the lining. The larger the car, the thicker and stronger this assembly is.

Shoes also have various holes and notches for attaching springs, pins, and struts. And many shoes have "nibs" or small indentations along the edge of the rim, which rest against shoe-support ledges on the backing plates. This arrangement keeps the shoes from dragging against the backing plate.

Though all its attachments help to hold the shoe in place, the holddown pin and spring serve no other function. And, just as the shoe must be held firmly in place, the brake-shoe lining must be solidly attached to the shoe.

Brake Lining Construction

Tremendous friction is generated when the brake-shoe lining is forced against the drum surface. This friction *is* the braking action. And, as mentioned previously, more than 1000 horsepower must be generated in this manner to fully stop a 4000 lb car from 60 mph within 300 ft. Thus, it readily becomes apparent that brakes *must* be much more efficient than other components of your car, such as the engine. To withstand the tremendous heat and force of thousands of stops, a brake lining must be a "tough cookie." It must be constructed well and must be attached properly to the shoe.

Brake linings are made of an asbestos compound, sometimes with a metallic substance included to increase resistance to wear and provide greater heat dissipation. Because of the tremendous heat generated in heavy braking operations, any lesser substance than asbestos might be charred or at least hardened. Asbestos tends to retain its

51

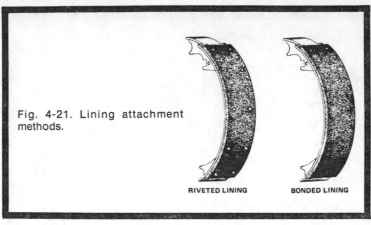

Fig. 4-21. Lining attachment methods.

RIVETED LINING   BONDED LINING

desirable characteristics in the face of such abuse. Though it wears well, it does wear away eventually. And it can be rendered less effective by foreign matter, such as hydraulic fluid, oils or greases, or other liquid or solid materials on its surface.

Eventually, brake linings must be replaced—the miracle being that modern brake linings last as long as they do. But their longevity depends upon both their construction and the firmness with which they are attached to their shoes.

Two methods of attaching linings to shoes are in common use. Many years ago all linings were riveted, as shown in Fig. 4-21A. Then, after World War II, bonding (cementing) linings to shoes came into vogue (Fig. 4-21B). Recently, however, some manufacturers have gone back to using rivets—the reason being that occasionally a bonded lining can work loose in heavy braking.

The advantage of the bonded lining is greater life. The advantage of riveted linings is greater certainty. The disadvantage is, of course, those rivet heads—they limit the life of the lining and, if the lining is allowed to wear too far, the rivets can rub against and score the drum surface.

Linings, whether bonded or riveted, differ from shoe to shoe. The secondary shoe usually employs a full-length lining that is centered on the shoe (Fig. 4-22). But the primary lining, for duoservo brakes, is always shorter than the secondary, and may be positioned on the shoe in a way to best effect servo action. For lining-to-drum contact is what braking is all about.

Whether new or rebuilt, most brake linings sold today are preground—that is, they are ground off on the ends so that they are left somewhat thicker in the middle (Fig. 4-23). When, you apply your brakes and the shoes force the linings into the

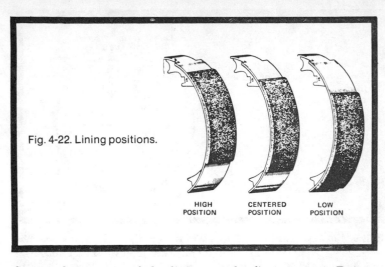

Fig. 4-22. Lining positions.

HIGH
POSITION

CENTERED
POSITION

LOW
POSITION

drums, the centers of the linings make first contact. But as brake pressure continues, both the shoe and the drum flex slightly, and the entire lining comes into contact with the drum (Fig. 4-24).

BRAKE DRUM SERVICING

No matter how or how well a lining is made, it must come into contact with and rub against a good drum surface. The drum surface must also withstand the high pressures without excessive flexing, and wear well in spite of the constant abrasion and terrific heat generated during brake applications.

Most drums are similar in general appearance (Fig. 4-25) and, for cars and light trucks, are very nearly the same size.

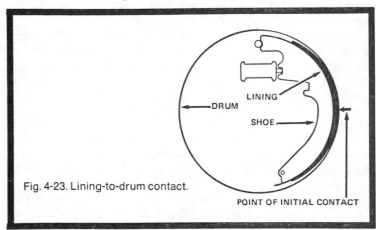

LINING

DRUM

SHOE

Fig. 4-23. Lining-to-drum contact.

POINT OF INITIAL CONTACT

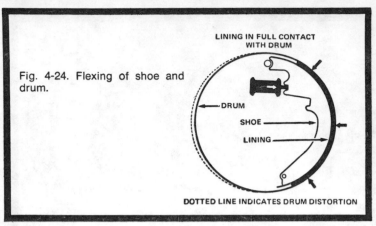

Fig. 4-24. Flexing of shoe and drum.

LINING IN FULL CONTACT WITH DRUM

DRUM

SHOE

LINING

DOTTED LINE INDICATES DRUM DISTORTION

Since 1952, manufacturers have made drums in diameters measured in increments of even half-inches (such as 11.000 in., 11.500 in., or 12.000 in.), up to 12.500 in. Also, except as noted in Table 4-1 drums are manufactured deliberately with a *rebore limit* of 0.060 in. That is, if a drum becomes worn or damaged, as illustrated in Fig. 4-26, and has to be turned on a lathe to bring it back into "round" or to smooth its inner surface, the total inside diameter can only be increased in this way by 0.060 in. This limit is not to be confused with the drum's *wear limit*.

Starting with 1971 models, all brake drums have a maximum-diameter dimension stamped into their casting. This number represents the drum's maximum wear limit and must not be confused with its rebore limit. For example, a drum with an original inner diameter of 11.000 in. may be rebored to only 11.060 in. but may be allowed to wear to 11.090 in. If 0.060 in. rebore does not correct or eliminate a drum's defects, *discard it*. Similarly, when a drum has worn to its wear limit (usually 0.090 in.), throw it away.

It should be noted that no more than 0.020 in. difference should exist between the diameters of drums opposite each other on an axle. Front- and rear-wheel differences are not critical, however.   All brake drums are also mounted in a universally similar manner. Rear-wheel drums usually are attached to the rear wheel and axle hub by the same lug bolts and nuts that hold the wheel in place. After the wheel is removed, if the parking brake is not engaged, the drum can be slipped off the brake assembly—sometimes it needs a few gentle taps to free it from the shoulder of the hub on which it rests. Front-wheel drums may be rigidly mounted.

Front-wheel drums normally may be removed with the wheel. Remove the cotter pin (Fig. 4-26), then the wheel nut

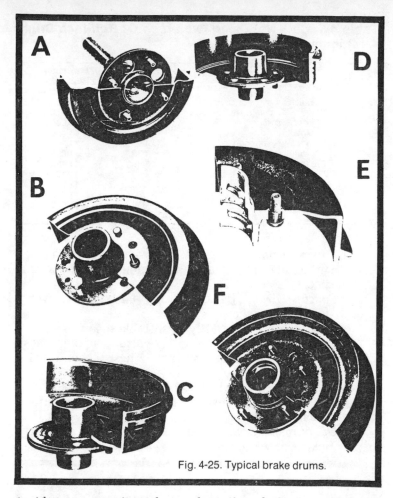

Fig. 4-25. Typical brake drums.

(nut loosens opposite to forward rotation of wheel), and grease retainer. Wiggle the wheel to loosen the outer bearing, then slide it off the axle. Remove the wheel and inner bearing. If drums require turning or other special attention, separate them from wheels by removing the lug nuts.

Some of the common problems involving drums are illustrated in Fig. 4-27. These include:

- *Scored drums*: Caused by hard foreign matter between lining and drum, or by excessively worn linings that allow rivets or shoe metal to rub against the drum surface.
- *Bell-mouthed drums*: Caused by excessive braking with a hot drum. Drum flexes beyond its elastic limits.

Table 4-1. Cars Not Using Even-Inch Or Half-Inch Drums As Original Equipment.

Year	Model	Original Drums Diameter (in.)	Maximum Rebore Diameter (in.)
*1956-70	Buick	*11.997"-12.022"	*12.080"
1961-62	Ford, Mercury & Thunderbird	*11.030"	*11.060"
1963-74	Ford & Mercury	11.030"	11.090"
1960-62	Lincoln	11.060"	11.120"
1963-66	Lincoln & Thunderbird	11.090"	11.150"
1967-71	Thunderbird	11.030"	11.090"
1967-68	Lincoln	11.090"	11.150"
*1969-70	Lincoln Continental	*11.090"	*11.130"
1971-73	Lincoln Continental	11.030"	11.090"
1968-72	Lincoln Mark III & IV	11.030"	11.090"
1973-74	Lincoln Mark IV	11.030"	11.090"

*Denotes exceptions to .060 in. rebore limit.

- *Tapered drum*: Caused by excessive wear along inner surface of drum, usually attributable to a loose or bent shoe.
- *Concave drum*: Caused by an unstable or overly flexible brake shoe.
- *Convex drum*: Caused by an unstable shoe.
- *Hard or chill spots*: Caused by variations in hardness of the drum's inner surface. These spots usually appear as shiny bluish or grayish blotches.
- *Heat-checked drum*: Caused by continued abuse to hard spots. They usually appear as minute surface lesions (cracks).
- *Out-of-round drum*: Caused by weakness or overly flexible area in drum.

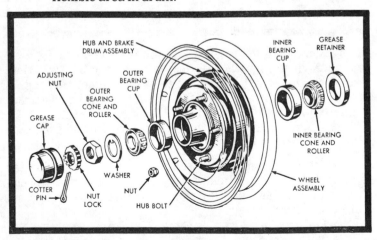

Fig. 4-26. Exploded view of front-wheel assembly.

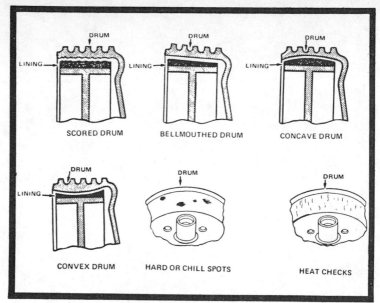

Fig. 4-27. Brake drum defects.

It should be noted, however, that hardly any of the above problems are likely to occur if your brakes are otherwise serviced when the need arises. *Take care of your brakes and they will take care of you.*

Disc Brake Fundamentals

5

In appearance, disc-brake assemblies bear almost no resemblance to drum-brake assemblies. Yet the principles upon which both operate are virtually identical. Both employ friction to effect brake action, and both press smooth asbestos pads against smooth metal surfaces to achieve this friction; but overall, disc brakes are more efficient and less complex.

As noted previously, the first American-built car to employ disc brakes was the hardly remembered Crosley, built in 1950. The first major user was Studebaker. And, while the Studebaker cars failed to "make it" with the American car-buying public, they did have a loyal following among truly appreciative automotive servicemen. Studebakers were superbly engineered (for the most part), efficient, and easy to service and maintain. Many of those Studebakers built with front-wheel disc brakes (1963—1966) may still be awaiting their first brake relinings. A 1963 Studebaker Hawk owned by one of the authors of this book could have gone farther than the 120,000 miles it did go before having its brakes relined. And another, a 1963 Avanti, was traded away with many miles on the speedometer, yet with plenty of wear left in the brake linings. This discussion, however, is not an endorsement of Studebaker, but rather a factual statement of what excellent instruments brakes have become on modern cars.

Since Studebaker's introduction of front-wheel disc brakes in 1963, coupled with rear-wheel drum brakes, this arrangement has become the norm on American cars. So much so, in fact, that cars without front-wheel disc brakes have become the minority. Thus, any brake discussion must deal as fully with disc brakes as with drum brakes.

Of all American-built cars, only Corvette (since 1966) and a few models of the Camaro offered four-wheel disc brakes prior to 1975. Then Chrysler introduced four-wheel disc brakes on its mammoth, the Imperial. But, with the present deemphasis on gas-guzzling behemoths, brake design has stabilized. The drum-and-disc combination will be with us for a long time.

DISC BRAKE ADVANTAGES

According to Bendix, disc brakes offer three major advantages over conventioanl drum brakes:

1. *Resistance to heat fade.* Disc brakes are generally more resistant to heat fade during high-speed or repeated braking because the design of disc brake rotors exposes more surface directly to the air, thus dissipating heat much more efficiently.
2. *Resistance to water fade.* Disc brakes are much less affected by water fade because, rotation of the rotor tends to throw off water rapidly, and the squeegee action of the pads pressed on the rotor tends to rapidly eliminate moisture from the braking surface.
3. *Better directional control.* Due to the clamping action of pads on rotors, disc brakes are less likely to pull, and generally produce a smoother, straighter stop.

DISC BRAKE MAKEUP

Four major disc brake manufacturers offer a variety of disc brakes for use on the many different car models produced by the Big Four manufacturers (Table 5-1), but there are only two basic disc brake designs. And each disc brake, regardless of its design, is made up of two basic assemblies:

- *Rotor assembly,* which is attached to and rotates with the wheel hub.
- *Caliper assembly,* which includes two pads and two plates, and one, two, or four hydraulic cylinder-piston assemblies.

Rotor Assembly

The rotor or disc is fixed to the wheel and turns with it. Both sides of the rotor are machined to serve as contact surfaces for the pads. After some experimentation by Briggs Cunningham and others with water cooling, engineers have settled upon air-cooled rotors. Those intended for light, low-performance vehicles are usually solid; more sophisticated designs feature radial vents for better heat transfer.

Caliper Assembly

The caliper assembly (Fig. 5-1) is a rugged metal casting or two castings bolted together, that contain the working parts

Table 5-1. Brake Applications.

Vehicle Make	Year	Type Brake	Brake Design	
Ambassador	1958-61	Drum	Bendix Duoservo	
(American	1962	Drum	Bendix Duoservo	
Motors)	1963-71	Drum	Bendix Duoservo	
	1965-68	Disc	Bendix	(F)
			Bendix Nonservo	(R)
	1969-70	Disc	Bendix	(F)
			Bendix Duoservo	(R)
	1971-74	Disc	Kelsey-Hayes	(F)
			Bendix Duoservo	(R)
American	1958-60	Drum	Wagner Servo Compound	
(Rambler)	1959	Drum	Bendix Nonservo	
	1960-61	Drum	Bendix Duoservo	
	1962	Drum	Bendix Duoservo	
	1963-68	Drum	Bendix Duoservo	
	1965-66	Drum	Wagner Servo Compound	
	1967	Disc	Bendix	(F)
			Bendix Nonservo	(R)
	1968	Disc	Bendix	(F)
			Bendix Duoservo	(R)
AMX (American Motors)	1969-71	Disc	Bendix	(F)
			Bendix Duoservo	(R)
Barracuda	1964-66	Drum	Bendix Duoservo	
(Plymouth)	1965-66	Disc	Kelsey-Hayes	(F)
			Bendix Duoservo	(R)
	1967-68	Drum	Bendix Duoservo	
	1967-69	Disc	Kelsey-Hayes	(F)
			Bendix Duoservo	(R)
	1969-74	Drum	Bendix Duoservo	
	1970-74	Disc	Kelsey-Hayes	(F)
			Bendix Duoservo	(R)
Belvedere	1957-61	Drum	Center Plane	
(Plymouth)	1962-66	Drum	Bendix Duoservo	
	1966	Disc	Bendix	(F)
			Bendix Duoservo	(R)
	1967-68	Drum	Bendix Duoservo	
	1967-69	Disc	Bendix	(F)
			Bendix Duoservo	(R)
	1969-71	Drum	Bendix Duoservo	
	1970-71	Disc	Kelsey-Hayes	(F)
			Bendix Duoservo	(R)
Buick	1950-62	Drum	Bendix Duoservo	
(LeSabre Invicta,	1963-66	Drum	Bendix Duoservo	
	1967-70	Drum	Bendix Duoservo	
Wildcat, Electra)	1967-69	Disc	Bendix	(F)
			Bendix Duoservo	(R)
	1970-74	Disc	Delco-Moraine	(F)
			Bendix Duoservo	(R)
Cadillac	1953-61	Drum	Bendix Duoservo	
	1962	Drum	Bendix Duoservo	

Vehicle Make	Year	Type Brake	Brake Design	
	1963	Drum	Bendix Duoservo	
	1964-68	Drum	Bendix Duoservo	
	1968-74	Disc	Delco-Moraine	(F)
			Bendix Duoservo	(R)
Camaro (Chevrolet)	1967-70	Drum	Bendix Duoservo	
	1967-68	Disc	Delco-Moraine	(F)
			Bendix Duoservo	(R)
	1969-74	Disc	Delco Moraine	(F)
			Bendix Duoservo	(R)
Challenger (Dodge)	1970-71	Drum	Bendix Duoservo	
	1970-74	Disc	Kelsey-Hayes	(F)
			Bendix Duoservo	(R)
Charger (Dodge)	1966	Drum	Bendix Duoservo	
	1966	Disc	Bendix	(F)
			Bendix Duoservo	(R)
	1967-68	Drum	Bendix Duoservo	
	1967-69	Disc	Bendix	(F)
			Bendix Duoservo	(R)
	1969-71	Drum	Bendix Duoservo	
	1970-74	Disc	Kelsey-Hayes	(F)
			Bendix Duoservo	(R)
Chevelle	1964-66	Drum	Bendix Duoservo	
	1967-71	Drum	Bendix Duoservo	
	1967-68	Disc	Delco-Moraine	(F)
			Bendix Duoservo	(R)
	1969-74	Disc	Delco-Moraine	(F)
			Bendix Duoservo	(R)
Cheverolet	1951-62	Drum	Bendix Duoservo	
	1963-66	Drum	Bendix Duoservo	
	1967-70	Drum	Bendix Duoservo	
	1967-68	Disc	Delco-Moraine	(F)
			Bendix Duoservo	(R)
	1969-74	Disc	Delco-Moraine	(F)
			Bendix Duoservo	(R)
Chevy II and Nova	1962	Drum	Bendix Duoservo	
	1963-66	Drum	Bendix Duoservo	
	1967-71	Drum	Bendix Duoservo	
	1967-68	Disc	Delco-Moraine	(F)
			Bendix Duoservo	(R)
	1969-74	Disc	Delco-Moraine	(F)
			Bendix Duoservo	(R)
Chrysler	1958-62	Drum	Center Plane	
	1963-66	Drum	Bendix Duoservo	
	1966	Disc	Budd	(F)
			Bendix Duoservo	(R)
	1967-68	Drum	Bendix Duoservo	
	1967-68	Disc	Budd	(F)
			Bendix Duoservo	(R)
	1969-71	Drum	Bendix Duoservo	
	1969-73	Disc	Kelsey-Hayes	(F)

Vehicle Make	Year	Type Brake	Brake Design	
			Bendix Duoservo	(R)
	1974	Disc	Sliding Caliper	(F)
			Bendix Duoservo	(R)
Classic (American Motors)	1958-61	Drum	Wagner Servo Compound	
	Ea. '62	Drum	Wagner Servo Compound	
	Lt. '62	Drum	Wagner Servo Compound	
	1963-66	Drum	Wagner Servo Compound	
	1964-66	Drum	Bendix Duoservo	
	1964-66	Disc	Bendix	(F)
			Bendix Nonservo	(R)
Comet (Mercury)	1960-62	Drum	Bendix Duoservo	
	1963-66	Drum	Bendix Duoservo	
	1967-71	Drum	Bendix Duoservo	
	1967	Disc	Kelsey-Hayes	(F)
			Bendix Duoservo	(R)
	1968-74	Disc	Kelsey-Hayes	(F)
			Bendix Duoservo	(R)
Coronet (Dodge)	1957-61	Drum	Center Plane	
	1962-66	Drum	Bendix Duoservo	
	1966	Disc	Bendix	(F)
			Bendix Duoservo	(R)
	1967-68	Drum	Bendix Duoservo	
	1967-69	Disc	Bendix	(F)
			Bendix Duoservo	(R)
	1969-71	Drum	Bendix Duoservo	
	1970-74	Disc	Kelsey-Hayes	(F)
			Bendix Duoservo	(R)
Corvair	1960-62	Drum	Bendix Duoservo	
	1963-66	Drum	Bendix Duoservo	
	1967-69	Drum	Bendix Duoservo	
Corvette	1954-62	Drum	Bendix Duoservo	
	1963-64	Drum	Bendix Duoservo	
	1965-66	Disc	Delco-Moraine	
	1967-74	Disc	Delco-Moraine	
Cougar (Mercury)	1967-71	Drum	Bendix Duoservo	
	1967	Disc	Kelsey-Hayes	(F)
			Bendix Duoservo	(R)
	1968-74	Disc	Floating Caliper	(F)
			Bendix Duoservo	(R)

Cuda — See Valiant

Cutlass — See F-85

Vehicle Make	Year	Type Brake	Brake Design	
Dart (Dodge)	1960-61	Drum	Center Plane	
	1962-66	Drum	Bendix Duoservo	
	1965-66	Disc	Kelsey-Hayes	(F)
			Bendix Duoservo	(R)
	1967-68	Drum	Bendix Duoservo	

Vehicle Make	Year	Type Brake	Brake Design	
	1967-72	Disc	Kelsey-Hayes	(F)
			Bendix Duoservo	(R)
	1969-72	Drum	Bendix Duoservo	
DeSoto	1957-61	Drum	Center Plane	
Dodge	1957-61	Drum	Center Plane	
(Monaco and	1962-66	Drum	Bendix Duoservo	
Polaro)	1965-66	Disc	Budd	(F)
			Bendix Duoservo	(R)
	1967-68	Drum	Bendix Duoservo	
	1967-68	Disc	Budd	(F)
			Bendix Duoservo	(R)
	1969-71	Drum	Bendix Duoservo	
	1969-73	Disc	Kelsey-Hayes	(F)
			Bendix Duoservo	(R)
Duster — See Valiant				
Eldorado	1967	Drum	Bendix Duoservo	
(Cadillac)	1967-68	Disc	Kelsey-Hayes	(F)
			Bendix Duoservo	(R)
	1969-74	Disc	Delco-Moraine	(F)
			Bendix Duoservo	(R)
Edsel	1958-60	Drum	Bendix Duoservo	
F-85 and	1961-62	Drum	Bendix Duoservo	
Cutlass	1963-66	Drum	Bendix Duoservo	
(Oldsmobile)	1967-71	Drum	Bendix Duoservo	
	1967-68	Disc	Delco-Moraine	(F)
			Bendix Duoservo	(R)
	1969-74	Disc	Delco-Moraine	(F)
			Bendix Duoservo	(R)
Fairlane	1962-66	Drum	Bendix Duoservo	
and Torino	1967-71	Drum	Bendix Duoservo	
	1967	Disc	Kelsey-Hayes	(F)
			Bendix Duoservo	(R)
	1968-74	Disc	Floating Caliper	(F)
			Bendix Duoservo	(R)
Falcon	1960-62	Drum	Bendix Duoservo	
	1963-66	Drum	Bendix Duoservo	
	1967-70	Drum	Bendix Duoservo	
	1967	Disc	Kelsey-Hayes	(F)
			Bendix Duoservo	(R)
	1968-70	Disc	Floating Caliper	(F)
			Bendix Duoservo	(R)
Firebird	1967-71	Drum	Bendix Duoservo	
(Pontiac)	1967-68	Disc	Delco-Moraine	(F)
			Bendix Duoservo	(R)
	1969-74	Disc	Delco-Moraine	(F)
			Bendix Duoservo	(R)
Ford	1951-60	Drum	Bendix Duoservo	
	1961-66	Drum	Bendix Duoservo	
	1966	Disc	Kelsey-Hayes	(F)

Vehicle Make	Year	Type Brake	Brake Design	
			Bendix Duoservo	(R)
	1967-71	Drum	Bendix Duoservo	
	1967	Disc	Kelsey-Hayes	(F)
			Bendix Duoservo	(R)
	1968-74	Disc	Floating Caliper	(F)
			Bendix Duoservo	(R)
Grand Prix	1969-70	Drum	Bendix Duoservo	
(Pontiac)	1969-74	Disc	Delco-Moraine	(F)
			Bendix Duoservo	(R)
Gremlin	1971	Drum	Bendix Duoservo	
(American	1972-74	Disc	Kelsey-Hayes	(F)
Motors			Bendix Nonservo	(R)
Hornet	1970-71	Drum	Bendix Duoservo	
(American	1970	Disc	Bendix	(F)
Motors)			Bendix Duoservo	(R)
	1971-74	Disc	Kelsey-Hayes	(F)
			Bendix Duoservo	(R)
Imperial	1958-62	Drum	Center Plane	
	1963-66	Drum	Bendix Duoservo	
	1966	Disc	Budd	(F)
			Bendix Duoservo	(R)
	1967-69	Disc	Budd	(F)
			Bendix Duoservo	(R)
	1970-74	Disc	Kelsey-Hayes	(F)
			Bendix Duoservo	(R)
Javelin	1968-71	Drum	Bendix Duoservo	
(American	1968-70	Disc	Bendix	(F)
Motors)			Bendix Duoservo	(R)
	1971-74	Disc	Kelsey-Hayes	(F)
			Bendix Duoservo	(R)
Lancer	1961-62	Drum	Bendix Duoservo	
Lincoln	1951-58	Drum	Bendix Duoservo	
Continental	1959-64	Drum	Bendix Duoservo	
	1965-66	Disc	Kelsey-Hayes	(F)
			Bendix Duoservo	(R)
	1967-69	Disc	Kelsey-Hayes	(F)
			Bendix Duoservo	(R)
	1970-71	Disc	Kelsey-Hayes	(F)
			Bendix Duoservo	(R)
	1972-74	Disc	Floating Caliper	(F)
			Bendix Duoservo	(R)
Lincoln Mark III	1958-71	Disc	Kelsey-Hayes	(F)
			Bendix Duoservo	(R)
Lincoln Mark IV	1972-74	Disc	Floating Caliper	(F)
			Bendix Duoservo	(R)
Marlin	1965-67	Drum	Bendix Duoservo	
(American	1965-67	Disc	Bendix	(F)
Motors)			Bendix Nonservo	(R)
Matador	1971-74	Drum	Bendix Duoservo	

Vehicle Make	Year	Type Brake	Brake Design	
(American Motors)	1971-74	Disc	Kelsey-Hayes	(F)
			Bendix Duoservo	(R)
Maverick	1969-74	Drum	Bendix Duoservo	
Mercury	1958-66	Drum	Bendix Duoservo	
	1966	Disc	Kelsey-Hayes	(F)
			Bendix Duoservo	(R)
	1967-71	Drum	Bendix Duoservo	
	1967	Disc	Kelsey-Hayes	(F)
			Bendix Duoservo	(R)
	1968-74	Disc	Floating Caliper	(F)
			Bendix Duoservo	(R)
Meteor (Mercury)	1962-63	Drum	Bendix Duoservo	
Monte Carlo (Chevrolet)	1970-74	Disc	Delco-Moraine	(F)
			Bendix Duoservo	(R)
Montego	1968-71	Drum	Bendix Duoservo	
	1968-74	Disc	Floating Caliper	(F)
			Bendix Duoservo	(R)
Mustang	1965-66	Drum	Bendix Duoservo	
	1965-66	Disc	Kelsey-Hayes	(F)
			Bendix Duoservo	(R)
	1967-71	Drum	Bendix Duoservo	
	1967	Disc	Kelsey-Hayes	(F)
			Bendix Duoservo	(R)
	1968-73	Disc	Floating Caliper	(F)
			Bendix Duoservo	(R)
Mustang II	1974	Disc	Floating Caliper	(F)
			Bendix Duoservo	(R)
Nova—See Chevy II				
Oldsmobile	1954-62	Drum	Bendix Duoservo	
	1963-66	Drum	Bendix Duoservo	
	1967-70	Drum	Bendix Duoservo	
	1967-68	Disc	Delco-Moraine	(F)
			Bendix Duoservo	(R)
	1969-71	Disc	Delco-Moraine	(F)
			Bendix Duservo	(R)
Pinto	1971-74	Drum	Bendix Duoservo	
	1971-73	Disc	Bendix	(F)
			Bendix Duoservo	(R)
	1974	Disc	Floating Caliper	(F)
			Bendix Duoservo	(R)
Plymouth (Fury)	1957-61	Drum	Center Plane	
	1962-66	Drum	Bendix Duoservo	
	1965-66	Disc	Budd	(F)
			Bendix Duoservo	(R)
	1967-68	Drum	Bendix Duoservo	
	1967-68	Disc	Budd	(F)
			Bendix Duoservo	(R)
	1969-73	Drum	Bendix Duoservo	

Vehicle Make	Year	Type Brake	Brake Design	
	1969-73	Disc	Kelsey-Hayes	(F)
			Bendix Duoservo	(R)
	1974	Disc	Floating Caliper	(F)
			Bendix Duoservo	(R)
Pontiac	1951-62	Drum	Bendix Duoservo	
	1963-66	Drum	Bendix Duoservo	
	1967-70	Drum	Bendix Duoservo	
	1967-68	Disc	Delco-Moraine	(F)
			Bendix Duoservo	(R)
	1969-74	Disc	Delco-Moraine	(F)
			Bendix Duoservo	(R)
Rebel (American Motors)	1967-69	Drum	Wagner Servo Compound	
	1967-70	Drum	Bendix Duoservo	
	1967-68	Disc	Bendix	(F)
			Bendix Nonservo	(R)
	1969-70	Disc	Bendix	(F)
			Bendix Duoservo	(R)
Riviera	1963-66	Drum	Bendix Duoservo	
	1967-70	Drum	Bendix Duoservo	
	1967-69	Disc	Bendix	(F)
			Bendix Duoservo	(R)
	1970-74	Disc	Delco-Moraine	(F)
			Bendix Duoservo	(R)
Road Runner — See Satellite				
Satellite and Road Runner (Plymouth)	1965-66	Drum	Bendix Duoservo	
	1966	Disc	Bendix	(F)
			Bendix Duoservo	(R)
	1967-68	Drum	Bendix Duoservo	
	1967-69	Disc	Bendix	(F)
			Bendix Duoservo	(R)
	1969-74	Drum	Bendix Duoservo	
	1970-74	Disc	Kelsey-Hayes	(F)
			Bendix Duoservo	(R)
Special and Skylark (Buick)	1961-62	Drum	Bendix Duoservo	
	1963-66	Drum	Bendix Duoservo	
	1967-71	Drum	Bendix Duoservo	
	1967-68	Disc	Delco-Moraine	(F)
			Bendix Duoservo	(R)
	1969-74	Disc	Delco-Moraine	(F)
			Bendix Duoservo	(R)
Studebaker	1954-62	Drum	Wagner Servo Compound	
	1963-66	Drum	Wagner Servo Compound	
	1963-66	Disc	Bendix	(F)
			Bendix Nonservo	(R)
Tempest (Includes LeMans and GTO)	1961-62	Drum	Bendix Duoservo	
	1963-66	Drum	Bendix Duoservo	
	1967-71	Drum	Bendix Duoservo	
	1967-68	Disc	Delco-Moraine	(F)

Vehicle Make	Year	Type Brake	Brake Design	
			Bendix Duoservo	(R)
	1969-74	Disc	Delco-Moraine	(F)
			Bendix Duoservo	(R)
Thunderbird	1955-58	Drum	Bendix Duoservo	
	1959-64	Drum	Bendix Duoservo	
	1965-66	Disc	Kelsey-Hayes	(F)
			Bendix Duoservo	(R)
	1967	Disc	Kelsey-Hayes	(F)
			Bendix Duoservo	(R)
	1968-74	Disc	Floating Caliper	(F)
			Bendix Duoservo	(R)
Torino — See Fairlane				
Toronado	1966	Drum	Bendix Duoservo	
(Oldsmobile)	1967-69	Drum	Bendix Duoservo	
	1967-68	Disc	Kelsey-Hayes	(F)
			Bendix Duoservo	(R)
	1969-74	Disc	Delco-Moraine	(F)
			Bendix Duoservo	(R)
Valiant	1960-66	Drum	Bendix Duoservo	
(Plymouth)	1965-66	Disc	Kelsey-Hayes	(F)
			Bendix Duoservo	(R)
	1967-68	Drum	Bendix Duoservo	
	1969-71	Drum	Bendix Duoservo	
	1967-72	Disc	Kelsey-Hayes	(F)
			Bendix Duoservo	(R)
	1973-74	Disc	Floating Caliper	(F)
			Bendix Duoservo	(R)
Vega	1971-74	Disc	Delco-Moraine	(F)
(Cheverolet)			Nonservo	(R)
Ventura	1971	Drum	Bendix Duoservo	
(Pontiac)	1971-74	Disc	Delco-Moraine	(F)
			Bendix Duoservo	(R)

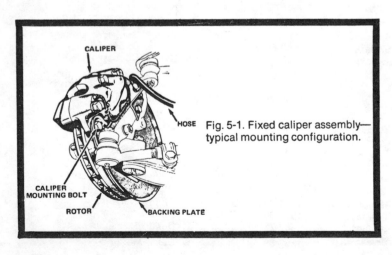

Fig. 5-1. Fixed caliper assembly—typical mounting configuration.

CALIPER

HOSE

CALIPER MOUNTING BOLT

ROTOR

BACKING PLATE

of the caliper assembly. In a fixed-caliper system, this assembly is attached rigidly to the front-wheel steering knuckle or to the backing plate. A cross section of a typical four-cylinder caliper assembly is shown in Fig. 5-2.

CALIPER CONFIGURATIONS

Caliper assemblies are classed as *fixed* or *floating*. In the fixed-caliper system, hydraulic pressure, transmitted through the master cylinder and the brake line, pushes the pistons in the caliper cylinders and forces the pads to clamp the rotor (Fig. 5-2). Friction from this clamping action slows or stops the rotor, which is rigidly attached to the wheel.

Then, as foot pressure is removed, hydraulic pressure against the pistons is released, and the clamping action ceases. Unlike drum brakes, however, no springs are needed to pull the brake pads away from the rotors. Tension within the pads or seals is adequate to pull them away from the rotor.

The floating (or sliding) caliper differs from the fixed caliper in design but not in principle. The significant advantage of this design (Figs. 5-3 and 5-4) is that it does with one cylinder what the *fixed* design accomplishes with two or four cylinders. A cross section (Fig. 5-5) illustrates how hydraulic pressure forces the primary pad against the rotor while, at the same time, the entire caliper slides in the opposite direction of piston movement to press the secondary pad against the rotor. A much more uniform clamping action is achieved with fewer moving (and wearing) parts. And the result is improved efficiency with reduced maintenance.

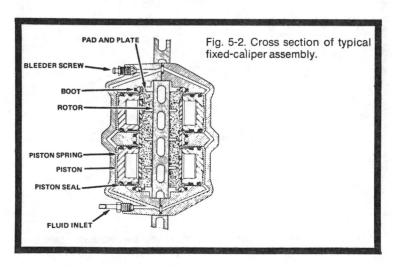

PAD AND PLATE

BLEEDER SCREW

BOOT

ROTOR

PISTON SPRING

PISTON

PISTON SEAL

FLUID INLET

Fig. 5-2. Cross section of typical fixed-caliper assembly.

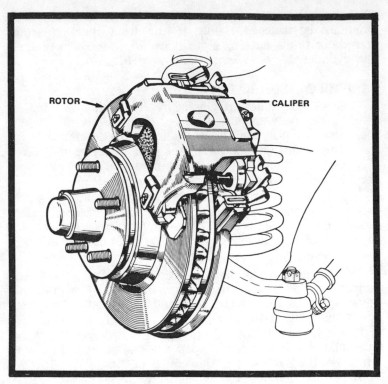

Fig. 5-3. Typical floating-caliper disc brake.

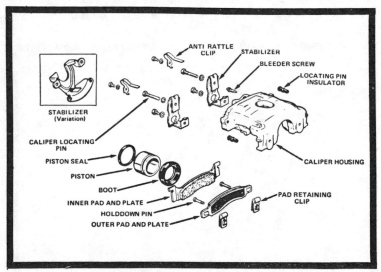

Fig. 5-4. Floating-caliper brake in exploded view.

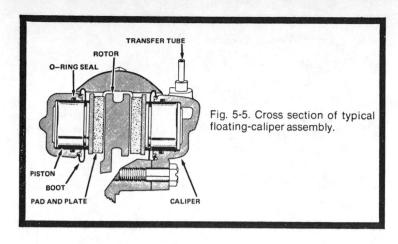

Fig. 5-5. Cross section of typical floating-caliper assembly.

Fixed Caliper Disc Brakes

Most early disc brakes used were of the fixed-caliper, four-piston design. Four major manufacturers provided these brakes, each employing only minor variations in design.

Bendix Four-Piston Caliper. The Bendix four-piston, fixed caliper disc brake (Fig. 5-6) was introduced in 1965 as optional equipment on the larger American Motors cars. Its major shortcoming, was its lack of ventilation—the rotor and the caliper were much more enclosed than an improved version

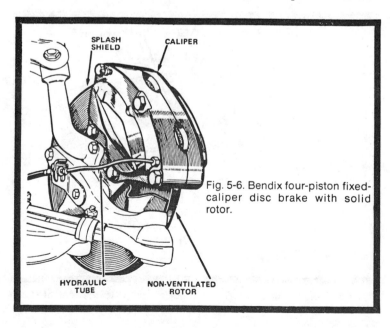

Fig. 5-6. Bendix four-piston fixed-caliper disc brake with solid rotor.

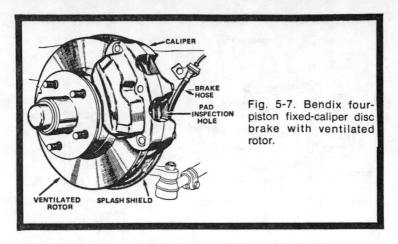

Fig. 5-7. Bendix four-piston fixed-caliper disc brake with ventilated rotor.

(Fig. 5-7) of the same brake, introduced on 1966 Plymouths and Dodges and, in 1967, on Buicks. This improved brake also made use of a ventilated rotor.

Not all disc brakes mount in precisely the same position. For example, the caliper shown in Fig. 5-6 is mounted forward of the front-wheel spindle and is attached to a bracket mounted on the steering knuckle. But the caliper shown in Fig. 5-8—the one used on those 1967 Plymouths and Dodges—is mounted to the rear of the front-wheel spindle and is secured directly to

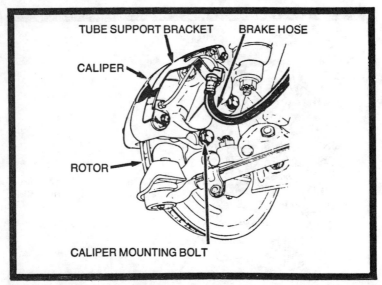

Fig. 5-8. Bendix four-piston fixed-caliper mounted to the rear of front-wheel spindle.

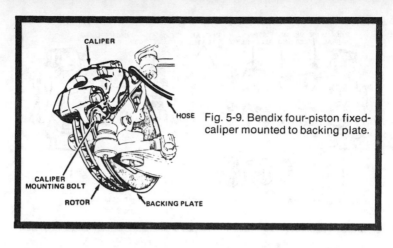

Fig. 5-9. Bendix four-piston fixed-caliper mounted to backing plate.

the steering knuckle. Finally, the caliper shown in Fig. 5-9—the one used on the 1967 Buicks—is attached directly to the backing plate, which also serves as a splash shield.

This Bendix caliper consists of two halves held together by three bolts. And each caliper half contains two separate cylinders and pistons (Fig. 5-10). Internal passages connect each cylinder to the hydraulic inlet and to the bleeder screw. Seals between the two halves prevent fluid leakage where these internal passages are joined.

All four pistons are alike: a lip-type seal (Fig. 5-11) fits into the groove near the bottom of each piston, and a piston spring serves to help keep the piston aligned in the cylinder.

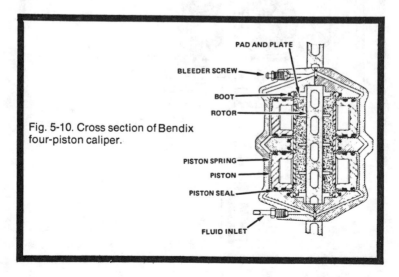

Fig. 5-10. Cross section of Bendix four-piston caliper.

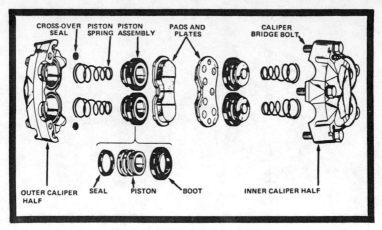

CROSS-OVER SEAL   PISTON SPRING   PISTON ASSEMBLY   PADS AND PLATES   CALIPER BRIDGE BOLT

OUTER CALIPER HALF   SEAL   PISTON   BOOT   INNER CALIPER HALF

Fig. 5-11. Exploded view of caliper.

Also, a dust boot, seated in a groove on the piston and against the adjoining caliper surface, protects the cylinder bore from dust and moisture.

A cross section of this Bendix brake is shown in the released position in Fig. 5-12, and in the applied position in Fig. 5-13.

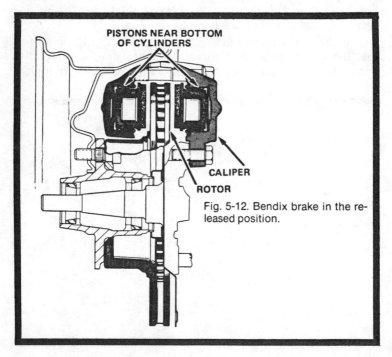

PISTONS NEAR BOTTOM OF CYLINDERS

CALIPER

ROTOR

Fig. 5-12. Bendix brake in the released position.

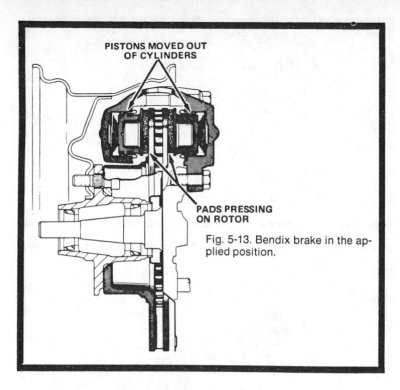

PISTONS MOVED OUT OF CYLINDERS

PADS PRESSING ON ROTOR

Fig. 5-13. Bendix brake in the applied position.

Budd Four-Piston Caliper. The Budd brake was introduced in 1965 as optional equipment on some Plymouth, Dodge, and Chrysler models through 1968. It differs only slightly from the Bendix model; however, it employed an antirattle spring (Fig. 5-14) to hold the pads and plates more firmly in place. An exploded view of this brake assembly is provided in Fig. 5-15.

Delco-Moraine Four-Piston Caliper. This brake, also introduced in 1965, was used on all four wheels of Corvettes and on some Camaro models. Also, it has been used on the front wheels of Chevrolets, Pontiacs, Oldsmobiles, and Buicks (Fig. 5-16). Two versions are shown in Figs. 5-17 and 5-18: the earlier version (Fig. 5-17), used only on 1965 and 1966 Corvettes, employed rather odd-shaped pistons and piston insulators. The later version (Fig. 5-18) had more conventional pistons, and the piston insulators were eliminated. A retaining pin serves much the same function as the antirattle spring used by the Budd caliper. An exploded view of this brake assembly is shown in Fig. 5-19.

Kelsey-Hayes Four-Piston Caliper. This disc brake was introduced on some Ford cars in 1965 and enjoyed widespread application by Ford until 1968. As can be seen in Fig. 5-20, it

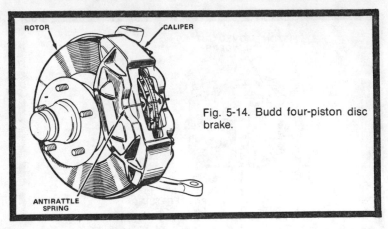

Fig. 5-14. Budd four-piston disc brake.

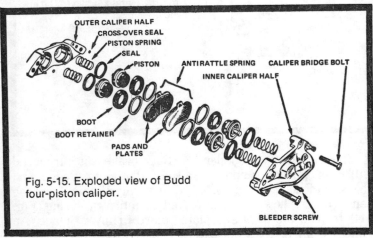

Fig. 5-15. Exploded view of Budd four-piston caliper.

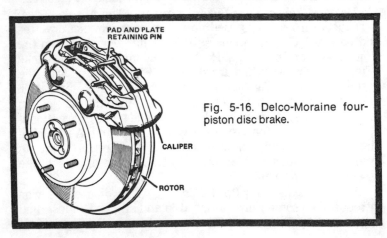

Fig. 5-16. Delco-Moraine four-piston disc brake.

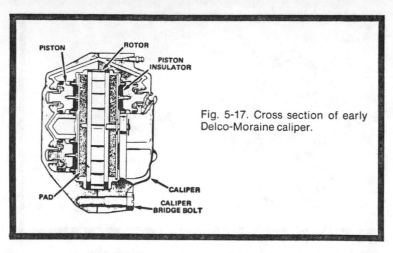

Fig. 5-17. Cross section of early Delco-Moraine caliper.

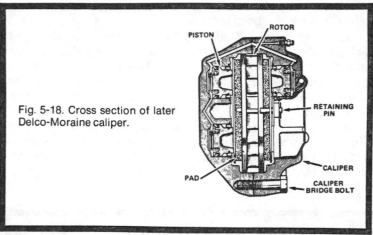

Fig. 5-18. Cross section of later Delco-Moraine caliper.

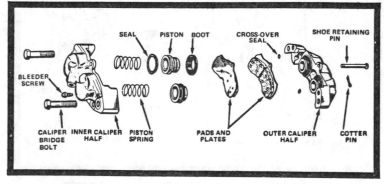

Fig. 5-19. Exploded view of Delco-Moraine four-piston caliper.

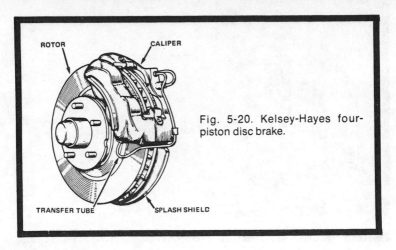

Fig. 5-20. Kelsey-Hayes four-piston disc brake.

differs very little from others of its class; however, its minor differences require brief mention.

As shown in Fig. 5-21, the piston seals are square-sectioned O-rings fitted into grooves machined into the cylinder bore rather than into the piston. Note also that there are no piston springs and there is no fluid crossover passage within the caliper; an external transfer tube provides fluid to the outer two cylinders.

Studebaker Two-Piston Fixed Caliper. This disc brake (Fig. 5-22) was employed by Studebaker in its pioneer applications (1963–1966). It featured bolt-on cylinders and an extended transfer tube. An exploded view is provided in Fig. 5-23.

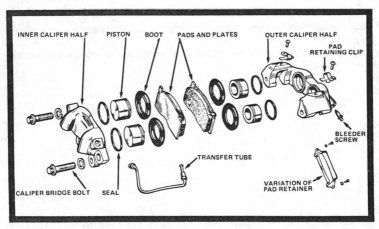

Fig. 5-21. Exploded view of Kelsey-Hayes caliper.

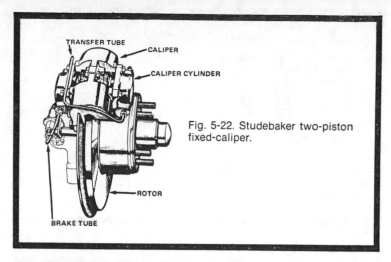

Fig. 5-22. Studebaker two-piston fixed-caliper.

Floating-Caliper Disc Brakes

The floating-caliper disc brake came into use on American cars in 1968. They are called "floating" because the calipers move sideways whenever the brakes are applied or released. Most consist of a single casting containing one cylinder and piston, though a few two-cylinder, two-piston calipers are in use.

In the floating-caliper disc brake the *O*-ring seal in the cylinder bore exerts heavy drag on the piston. Figure 5-24 shows a cross section of a floating caliper in its released position. The pads may touch the rotor lightly or, more likely,

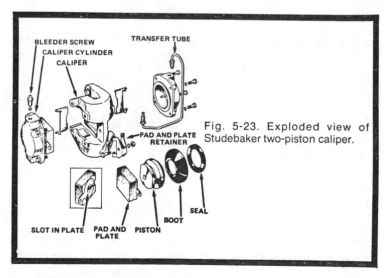

Fig. 5-23. Exploded view of Studebaker two-piston caliper.

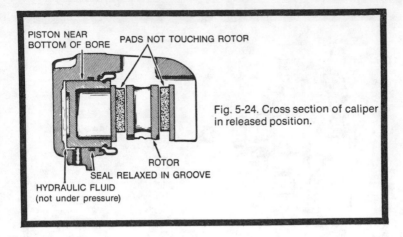

Fig. 5-24. Cross section of caliper in released position.

PISTON NEAR BOTTOM OF BORE

PADS NOT TOUCHING ROTOR

ROTOR

SEAL RELAXED IN GROOVE

HYDRAULIC FLUID (not under pressure)

there may be a small clearance between the pads and rotor. The seal is in a relaxed condition. When the brakes are applied, however, hydraulic pressure builds up in the caliper cylinder behind the piston and seal—and remember, hydraulic pressure is equal in all directions—pushing as hard against the rear of the cylinder as against the piston. Initially, however, the piston offers least resistance; therefore, it moves first. Then, as the piston presses the inner pad against the rotor, resistance to this movement is increased and the entire caliper is pushed in the opposite direction, pressing the outer pad against the caliper. As hydraulic pressure increases, both pads are forced more tightly against the rotor in a clamping action. This clamping action (Fig. 5-25) slows or stops rotation of the rotor and the wheel to which it is attached.

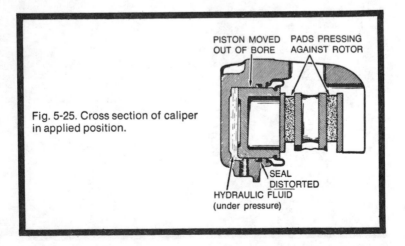

PISTON MOVED OUT OF BORE

PADS PRESSING AGAINST ROTOR

Fig. 5-25. Cross section of caliper in applied position.

SEAL DISTORTED

HYDRAULIC FLUID (under pressure)

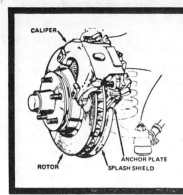

Fig. 5-26. Ford floating-caliper disc brake.

When the brakes are released, the piston seal returns to its normal position, pulling the piston back into the cylinder bore and releasing the pressure of the inner pad against the rotor. The caliper and outer pad return to their original position.

Kelsey-Hayes Floating Caliper. Kelsey-Hayes provides floating-caliper disc brakes for Ford Motor Company and many Chrysler cars.

The Ford Floating Caliper. The brake shown in Fig. 5-26, was introduced in 1968 on some Ford-built cars. The caliper on this brake consists of a one-piece casting containing a single cylinder and piston. A groove in the cylinder bore (Fig. 5-27) contains a square-sectioned *O*-ring seal, and—as on the Kelsey-Hayes fixed caliper—the piston must be inserted

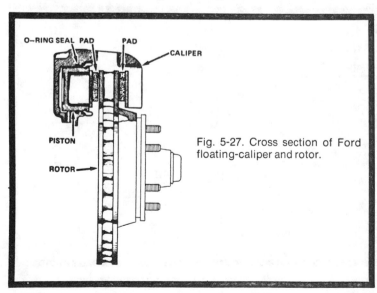

Fig. 5-27. Cross section of Ford floating-caliper and rotor.

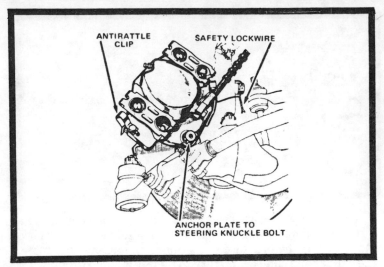

Fig. 5-28. Ford floating-caliper mounting.

through this seal. A boot is seated in the piston and against the caliper surface to protect the piston and cylinder bore from moisture and dust. And two threaded locating pins are used to attach the caliper assembly to the anchor plate, which is secured to the steering knuckle by two bolts (Fig. 5-28). Safety wire may also be used. An exploded view of this caliper assembly is shown in Fig. 5-29.

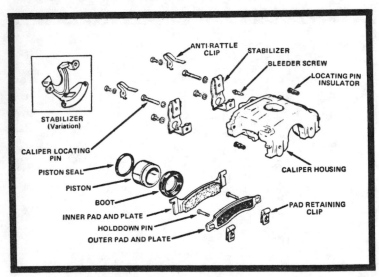

Fig. 5-29. Exploded view of Ford floating-caliper.

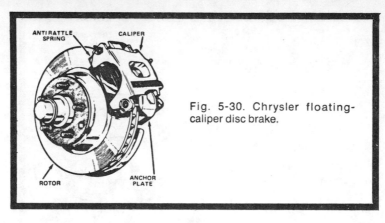

Fig. 5-30. Chrysler floating-caliper disc brake.

The Chrysler Floating Caliper. The Chrysler version, shown in Fig. 5-30, was introduced in 1969 on Plymouths, Dodges, and Chryslers. It also has a one-piece caliper casting containing a single hydraulic cylinder and piston. It is quite similar to the Ford caliper, except for two basic differences: The Chrysler caliper uses an antirattle spring to hold the outer plate and pad assembly in place. Also, it uses four rubber bushings as shown in Fig. 5-31.

In 1973, Chrysler introduced the *sliding-caliper* disc brake shown in Fig. 5-32. And in 1974, this design was used on all four wheels of the Imperial. It is basically the same as the brake discussed above except for the method of retaining the caliper in the adapter. An exploded view is shown in Fig. 5-33.

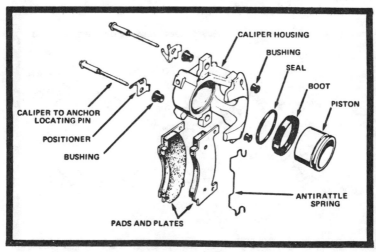

Fig. 5-31. Exploded view of Chrysler floating-caliper.

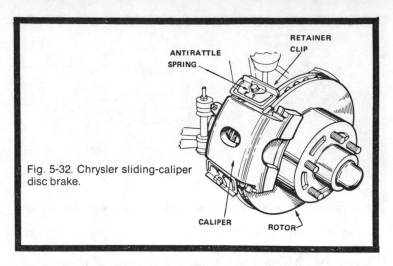

Fig. 5-32. Chrysler sliding-caliper disc brake.

Delco-Moraine Floating Caliper. This floating-caliper disc brake (Fig. 5-34) was introduced in 1968 on the front wheels of all Cadillac cars. It uses a one-piece casting containing a single cylinder and piston (Fig. 5-35). The brake shown has an anchor plate attached to the steering knuckle by two bolts; however, on later models, the anchor plate and steering knuckle are cast as one piece. An exploded view of this caliper is shown in Fig. 5-36.

Bendix CA Floating Caliper. CA stands for "center abutment," which is another way of getting at what happens in all these floating-caliper designs. This Bendix brake (Fig. 5-37) was introduced on the Ford *Pinto* in 1971. It is slightly less

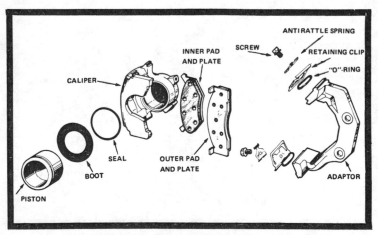

Fig. 5-33. Exploded view of Chrysler sliding caliper.

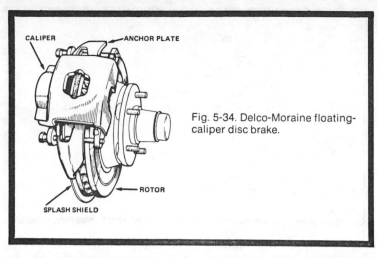

Fig. 5-34. Delco-Moraine floating-caliper disc brake.

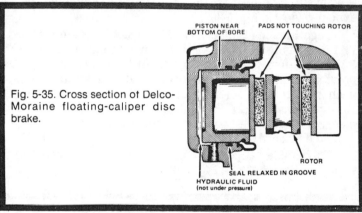

Fig. 5-35. Cross section of Delco-Moraine floating-caliper disc brake.

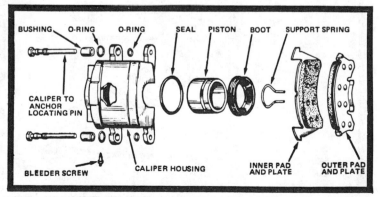

Fig. 5-36. Exploded view of Delco-Moraine floating-caliper disc brake.

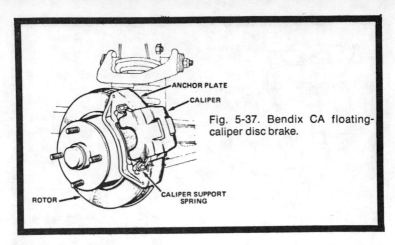

Fig. 5-37. Bendix CA floating-caliper disc brake.

complex, perhaps, than the floating calipers discussed previously, and it has fewer parts, as indicated by the exploded view in Fig. 5-38.

Vega Floating Caliper. The Chevrolet *Vega*, introduced in 1971, began life with a front-wheel disc brake of its own (Fig. 5-39). It differs little in design from those brakes discussed previously; however, take note of two design features: (1) the anchor plate and steering-knuckle casting is of one-piece construction and (2) the hub-and-rotor casting is a single piece. The rotor is nonventilated, and a splash shield is mounted inboard of the rotor.

The simple caliper assembly is further illustrated in the exploded view shown in Fig. 5-40.

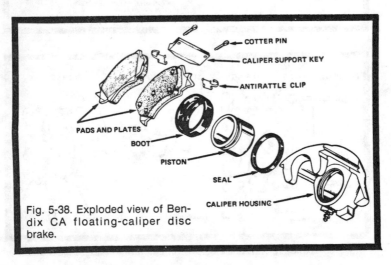

Fig. 5-38. Exploded view of Bendix CA floating-caliper disc brake.

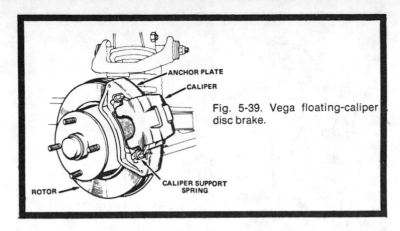

Fig. 5-39. Vega floating-caliper disc brake.

Dayton Floating Caliper. This design came into limited use in 1968 as optional equipment on the front wheel of ¾ and 1-ton Ford trucks. It has a two-piece caliper assembly (Fig. 5-44) that consists of the caliper and piston housing. The caliper, which contains two cylinders and pistons and two plates and pads, fits over the ventilated rotor and is mounted on the rear of the steering knuckle as shown in Fig. 5-42.

This two-cylinder caliper, shown in an exploded view in Fig. 5-43, operates in the same manner as the one-cylinder models discussed previously.

ROTORS

Starting in 1971, all disc brake rotors on American-made cars have a minimum-thickness dimension stamped on the

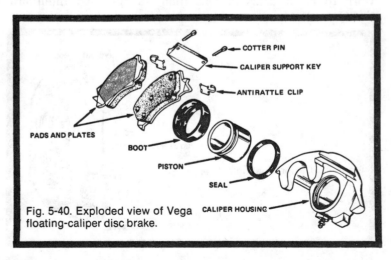

Fig. 5-40. Exploded view of Vega floating-caliper disc brake.

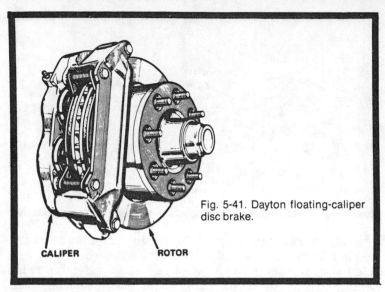

Fig. 5-41. Dayton floating-caliper disc brake.

CALIPER   ROTOR

casting. This dimension is the minimum wear thickness (more properly called the discard thickness) and *not* the minimum refinishing dimension. This latter dimension is 0.060 in. less than the rotor's original thickness. For example:

New rotor thickness ....1.290 in.
Minimum thickness after refinishing ......1.230 in.
Minimum wear or discard thickness ......1.215 in.

Never refinish a rotor to less than 0.015 in. of its minimum wear dimension. And *never* reinstall or continue to use a rotor worn to a thickness of less than its specified minimum thickness.

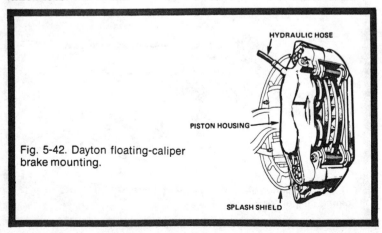

HYDRAULIC HOSE

PISTON HOUSING

Fig. 5-42. Dayton floating-caliper brake mounting.

SPLASH SHIELD

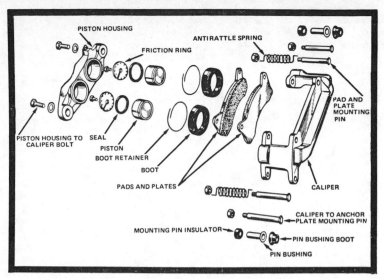

Fig. 5-43. Exploded view of Dayton floating-caliper disc brake.

Table 5-2 indicates minimum thickness for all American-built rotors through 1971.

Table 5-2. Disc-Brake Rotor Specifications.

Make	Year	Brake Design & Number of Pistons	Rotor Specifications (in.)		
			Thickness (New)	Minimum Refinishing Thickness	Thickness Variation (Parallelism)
American Motors	1965-70	Bendix-4	.500	.465	.0005
	1971	KH-1	1.000	.955	.0005
Barracuda	1965-69	KH-4	.810	.790	.0005
	1970	KH-1	1.005	.980	.0005
	1971	KH-1	1.005	.955	.0005
Belvedere	1966-69	Bendix-4	.885	.855	.0005
	1970	KH-1	1.005	.980	.0006
Buick	1967-69	Bendix-4	1.000	.980	.0005
	1970	DM-1	1.290	1.230	.0005
	1971	DM-1	1.290	1.230	.0005
Buick Special, Skylark	1967-68	DM-4	1.000	.980	.0005
	1969	DM-1	1.000	.965	.0005
	1970	DM-1	1.040	.980	.0005
	1971	DM-1	1.040	.980	.0005
Cadillac	1968-71	DM-1	1.250	1.230	.0007
Cadillac Eldorado	1967-68	KH-4	1.250	1.230	.0005
	1969-71	DM-1	1.210	1.190	.0005
Camaro	1967-68	DM-4	1.000	.980	.0005
	1969-70	DM-1	1.000	.980	.0005
	1971	DM-1	1.000	.980	.0005
Camaro Z-28[2]	1969	DM-4	1.250	1.230	.0005
Challenger	1970	KH-1	1.005	.980	.0005
	1971	KH-1	1.005	.955	.0005
Charger	1966-69	Bendix-4	.885	.855	.0005
	1970	KH-1	1.005	.980	.0005
	1971	KH-1	1.005	.955	.0005

Checker Motors	1969-70	DM-1	1.250	1.230	.0005
	1971	DM-1	1.280	1.230	.0005
Chevelle	1967-68	DM-4	1.000	.980	.0005
	1969-70	DM-1	1.000	.980	.0005
	1971	DM-1	1.000	.980	.0005
Chevrolet	1967-68	DM-4	1.250	1.230	.0005
	1969-70	DM-1	1.250	1.230	.0005
	1971	DM-1	1.250	1.230	.0005
Chevy II, Nova	1967-68	DM-4	1.000	.980	.0005
	1969-70	DM-1	1.000	.980	.0005
	1971	DM-1	1.000	.980	.0005
Chevrolet Truck 10, 20, 30 Series	1971	DM-1	1.250	1.230	.0005
Chrysler	1966-68	Budd-4	.875	.845	.0005
	1969-71	KH-1	1.245	1.195	.0005
Comet, Montego	1967	KH-4	.810	.795	.0007
	1968-71	KH-1	.935	.890	.0007
Coronet	1966-69	Bendix-4	.885	.855	.0005
	1970	KH-1	1.005	.980	.0005
	1971	KH-1	1.005	.955	.0005
Corvette[2]	1965-71	DM-4	1.250	1.230	.0005
Cougar	1967	KH-4	.810	.795	.0007
	1968-71	KH-1	.935	.890	.0007
Cutlass — See Olds F-85 Dart	1965-71	KH-4	.810	.795	.0005
Dodge (Monaco, Polara)	1966-68	Budd-4	.875	.845	.0005
	1969-71	KH-1	1.245	1.195	.0005
Dodge Trucks B-100, B-200	1971	KH-1	1.245	1.195	.0005
Fairlane, Torino	1967	KH-4	810	.795	.0007
	1968-71	KH-1	.935	.890	.0007
Falcon	1967	KH-4	.810	.795	.0007
	1968-70	KH-1	.935	.890	.0007
Firebird	1967-68	DM-4	1.000	.980	.0005
	1969	DM-1	1.000	.975	.0007
	1970	DM-1	1.085	.980	.0007
Ford	1967	KH-4	1.245	1.230	.0007
	1968	KH-1	1.180	1.170	.0007
	1969-71	KH-1	1.180	1.135	.0007
Ford Trucks, F-250, F-350	1958-71	Dayton-2	1.000	.955	.0010
Grand Prix	1967-68	DM-4	1.250	1.230	.0005
	1969	DM-1	1.000	.975	.0007
	1970-71	DM-1	1.085	.980	.0007
Imperial	1967-69	Budd-4	.875	.845	.0005
	1970-71	KH-1	1.245	1.195	.0005
'.incoln	1965-69	KH-4	1.245	1.230	.0007
	1970-71	KH-1	1.180	1.135	.0007
Mark III	1968	KH-1	1.180	1.170	.0007
	1969-71	KH-1	1.180	1.135	.0007
Mercury	1965-67	KH-4	1.245	1.230	.0007
	1968	KH-1	1.180	1.170	.0007
	1969-71	KH-1	1.180	1.135	.0007
Monte Carlo	1970	DM-1	1.000	.980	.0005
	1971	DM-1	1.000	.980	.0005
Montego—See Comet Mustang	1965-67	KH-4	.810	.795	.0007
	1968-71	KH-1	.935	.890	.0007
Nova—See Chevy II Oldsmobile	1967-68	DM-4	1.250	1.230	0005

	1969-70	DM-1	1.250	1.230	.0005
	1971	DM-1	1.290	1.230	.0005
Olds F-85	1967-68	DM-4	1.000	.980	.0005
	1969	DM-1	1.000	.980	.0005
	1970-71	DM-1	1.040	.980	.0005
Oldsmobile Toronado	1967-68	KH-4	1.245	1.230	.0005
	1969	DM-1	1.250	1.230	.0005
	1970	DM-1	1.205	1.185	.0005
	1971	DM-1	1.245	1.185	.0005
Plymouth (Fury)	1966-68	Budd-4	.875	.845	.0005
	1969-71	KH-1	1.245	1.195	.0005
Pontiac	1967-68	DM-4	1.250	1.230	.0005
	1969	DM-1	1.240	1.210	.0007
	1970-71	DM-1	1.250	1.230	.0007
Pinto	1971	Bendix-1	.750	.700	.0010
Riviera	1967-69	Bendix-4	1.000	.980	.0005
	1970	DM-1	1.290	1.230	.0005
	1971	DM-1	1.290	1.230	.0005
Satellite	1966-69	Bendix-4	.885	.855	.0005
	1970	KH-1	1.005	.980	.0005
	1971	KH-1	1.005	.955	.0005
Skylark-See Buick Special					
Tempest, Lemans	1967-68	DM-4	1.000	.980	.0005
	1969	DM-1	1.000	.975	.0007
	1970-71	DM-1	1.085	.980	.0007
Thunderbird	1965-67	KH-4	1.245	1.230	.0007
	1968	KH-1	1.180	1.170	.0007
	1969-71	KH-1	1.180	1.135	.0007
Torino—See Fairlane					
Valiant	1965-71	KH-4	.810	.795	.0005
Vega	1971	DM-1	.500	.455	.0005
Studebaker	1963-66	Bendix-2	.375	.345	.0005

[1] Minimum wear or discard thickness is approximately 0.015 in. less than refinishing thickness.

[2] Four-wheel disc brakes.

KH Kelsey-Hayes

DM Delco-Moraine

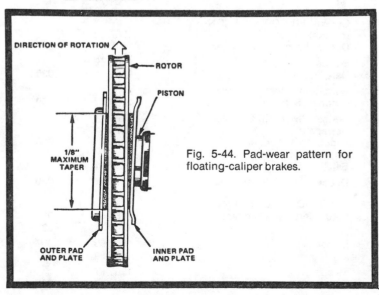

Fig. 5-44. Pad-wear pattern for floating-caliper brakes.

DIRECTION OF ROTATION

ROTOR

PISTON

1/8" MAXIMUM TAPER

OUTER PAD AND PLATE

INNER PAD AND PLATE

PAD WEAR

Disc pads used with floating calipers tend to wear in a taper pattern as shown in Fig. 5-44. This wear, caused by the manner in which the caliper is attached to its anchor, is normal and is no cause for concern unless the taper is excessive—more than ⅛ in. difference between top and bottom thickness. Even then, no corrective action is necessary, except to install a new pad-and-plate assembly.

Incidentally, always replace *all* pads and plates on both front-wheel brakes at the same time, and—if your car has four-wheel disc brakes—on both rear-wheel brakes at the same time. Otherwise you will experience uneven stops, and brake wear will be distorted.

Pad wear limits are shown in Table 5-3. This table is compiled from manufacturer's data and may not meet minimum legal requirements in all states. For example, Pennsylvania inspection requires that pads and linings have 40% of their original thickness.

Table 5-3. Minimum Disc-Brake Pad Thickness.

Model	Year	Minimum Pad (Lining) Thickness
AMERICAN MOTORS		
All	1965-70	.060[3]
All	1971	
CHECKER MOTORS		
All	1969-71	.125
CHRYSLER CORPORATION		
Imperial	1967-69	
Dodge, Plymouth, Chrysler	1966-68	.030
Barracuda	1965-69	
Dart, Valiant	1965-71	.030
Belvedere, Charger, Coronet, Satellite	1966-69	.030
Dodge, Plymouth, Chrysler	1969-71	
Belvedere	1970	
Imperial, Challenger,		.030
Charger, Coronet,		
Satellite, Barracuda	1970-71	
Dodge Trucks B-100, B-200	1971	.030
FORD MOTOR COMPANY		
Ford, Mercury, Thunderbird	1965-66	.030
	1967	.030
Lincoln	1965-66	.067
	1967-69	.067
Mustang	1965-67	
Comet, Cougar, Fairlane,		.030

Falcon	1967	
Ford, Mercury, Thunderbird, Mark III	1968-71	.030
Fairlane, Falcon	1968-70	
Mustang, Comet, Cougar, Montego, Torino	1968-71	.030
Pinto	1971	$.030^2$
F-250, F-350 trucks	1968-71	.067
GENERAL MOTORS CORPORATION Buick, Riviera	1967-69	.060
Buick Special, Skylark	1967-68	$.020^2$
Buick, Riviera	1970-71	
Buick Special, Skylark	1969-71	$.020^2$
Cadillac	1968-71	3
Cadillac Eldorado	1967-68	3
Cadillac Eldorado	1969-71	3
Chevrolet, Chevelle, Chevy II, Camaro	1967-68	$.030^2$
Chevrolet, Chevelle, Chevy II, Nova, Camaro	1969-71	$.020^2$
Monte Carlo	1970-71	
Corvette	1965-71	$.020^2$
Camaro Z-28	1969	
Chevrolet Trucks 10, 20, 30 Series	1971	$.033^2$
Oldsmobile, F-85	1967-68	.067
Oldsmobile, F-85, Cutlass	1969-71	3
Oldsmobile Toronado	1967-68	3
Oldsmobile Toronado	1969-71	3
Pontiac, Tempest Firebird	1967-68	$.020^2$
Pontiac, Tempest Firebird, Grand Prix	1969-71	.125
STUDEBAKER	1963-66	.200

[1]Four-wheel disc brakes.
[2]Over rivet heads.
[3]Approximately same thickness as plates

Troubleshooting

6

Troubleshooting brakes is not difficult. Find the symptom (or symptoms), read across the page to the probable cause, and take the suggested corrective action.

EXCESSIVE BRAKE-PEDAL "TRAVEL"

Brake Type	Probable Cause	Corrective Action
Drum	Excessive clearance between drums and linings.	Adjust brakes.
	Automatic adjusters not working.	Make several forward and reverse stops; if pedal docs not firm up, repair automatic adjusters.
	Bent or distorted shoe.	Replace shoes and linings in axle sets.
Drum or disc	Leak in hydraulic system.	Check (and replace or repair as necessary) master cylinder, wheel cylinders, calipers, tubes, and hoses. (See Table 6-1.) Repair or replace faulty parts.
	Air in hydraulic system.	Bleed hydraulic system.
	Poor-quality brake fluid (low boiling point).	Drain and flush hydraulic system, refill with new DOT-3 fluid.

Table 6-1. Hydraulic Leak Test.

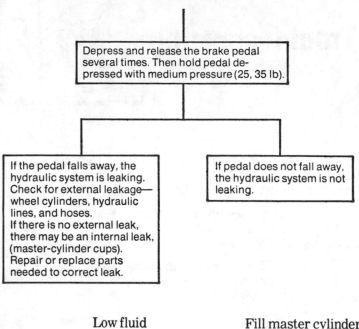

Depress and release the brake pedal several times. Then hold pedal depressed with medium pressure (25, 35 lb).

If the pedal falls away, the hydraulic system is leaking. Check for external leakage—wheel cylinders, hydraulic lines, and hoses. If there is no external leak, there may be an internal leak, (master-cylinder cups). Repair or replace parts needed to correct leak.

If pedal does not fall away, the hydraulic system is not leaking.

Low fluid level.	Fill master cylinder and bleed system.
Weak brake hoses that expand under pressure.	Replace defective hoses.
Improperly adjusted master-cylinder pushrod (manual or power booster).	Adjust pushrod.

INSUFFICIENT BRAKE-PEDAL TRAVEL,
BUT IMPROVES WITH DRIVING

Brake Type	Probable Cause	Corrective Action
Drum	Residual-pressure check valves not holding pressure in lines.	Repair or replace master cylinder.

Brake Type	Probable Cause	Corrective Action
Disc	Pad and plate "knockback" caused by loose wheel bearings or faulty front suspension.	Adjust, tighten, or replace loose or faulty parts.

BRAKE PEDAL "SPONGY"

Brake Type	Probable Cause	Corrective Action
Drum	Bent or distorted brake shoe.	Replace shoes and linings in axle sets.
	Shoes not centered in drums.	Adjust anchor pins.
	Cracked or thin drums.	Replace drums in axle sets.
Drum or Disc	Poor-quality brake fluid (low boiling point).	Drain and flush system; refill with new DOT-3 fluid; bleed system.
	Weak hoses that expand under pressure.	Replace defective hoses.
	Air in hydraulic system.	Bleed system.

PEDAL REQUIRES EXCESSIVE PUSHING FORCE

Brake Type	Probable Cause	Corrective Action
Drum	Poor-quality linings.	Replace shoes and linings in axle sets.
	Grease-soaked linings.	Replace grease seal and, if necessary, shoes and linings in axle sets.
	Fluid-soaked linings.	Repair wheel cylinder; if necessary, replace shoes and linings in axle sets.
	Glazed linings.	Replace shoes and linings in axle sets.

	Damaged or distorted shoes.	Replace shoes and linings in axle sets.
	Scored, belled, or badly worn drums.	Refinish or replace drums in axle sets.
Disc	Grease-soaked pads.	Replace grease seal; if necessary, replace pads and plates in axle sets.
	Fluid-soaked pads.	Repair caliper; if necessary, replace pads and plates in axle sets.
Drum or disc	Clogged tubes or hoses.	Replace tubes or hoses as necessary.
	Clogged or "frozen" master cylinder.	Repair or replace master cylinder.
	Low vacuum supply to power-brake booster.[1]	Tune or repair engine to obtain adequate vacuum. (See Table 6-2.)
	Loose or leaking vacuum hose to power-brake booster.	Tighten clamps or replace hose as required. (See Table 6-2.)
	Defective power-brake check valve.	Replace check valve. (See Table 6-2.)
	Faulty power-brake vacuum booster.	Replace power section of vacuum booster. (See Table 6-2.)

BRAKES "GRAB"

Brake Type	Probable Cause	Corrective Action
Drum	Incorrect or distorted shoes.	Replace with correct shoes and linings in axle sets.

Table 6-2. Power Brake Checks.

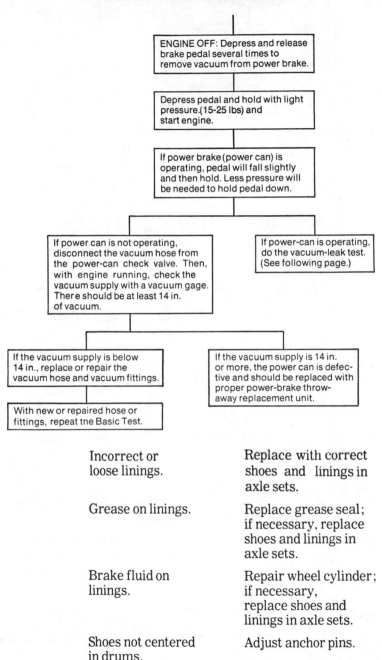

ENGINE OFF: Depress and release brake pedal several times to remove vacuum from power brake.

Depress pedal and hold with light pressure.(15-25 lbs) and start engine.

If power brake (power can) is operating, pedal will fall slightly and then hold. Less pressure will be needed to hold pedal down.

If power can is not operating, disconnect the vacuum hose from the power-can check valve. Then, with engine running, check the vacuum supply with a vacuum gage. There should be at least 14 in. of vacuum.

If power-can is operating, do the vacuum-leak test. (See following page.)

If the vacuum supply is below 14 in., replace or repair the vacuum hose and vacuum fittings.

If the vacuum supply is 14 in. or more, the power can is defective and should be replaced with proper power-brake throw-away replacement unit.

With new or repaired hose or fittings, repeat tne Basic Test.

Incorrect or loose linings.	Replace with correct shoes and linings in axle sets.
Grease on linings.	Replace grease seal; if necessary, replace shoes and linings in axle sets.
Brake fluid on linings.	Repair wheel cylinder; if necessary, replace shoes and linings in axle sets.
Shoes not centered in drums.	Adjust anchor pins.

Table 6-2. Cont'd.

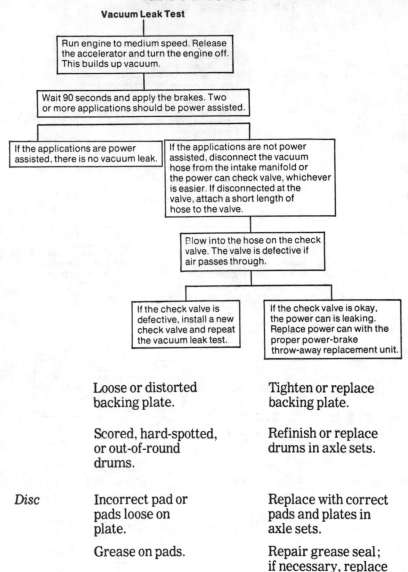

Vacuum Leak Test

Run engine to medium speed. Release the accelerator and turn the engine off. This builds up vacuum.

Wait 90 seconds and apply the brakes. Two or more applications should be power assisted.

If the applications are power assisted, there is no vacuum leak.

If the applications are not power assisted, disconnect the vacuum hose from the intake manifold or the power can check valve, whichever is easier. If disconnected at the valve, attach a short length of hose to the valve.

Blow into the hose on the check valve. The valve is defective if air passes through.

If the check valve is defective, install a new check valve and repeat the vacuum leak test.

If the check valve is okay, the power can is leaking. Replace power can with the proper power-brake throw-away replacement unit.

	Loose or distorted backing plate.	Tighten or replace backing plate.
	Scored, hard-spotted, or out-of-round drums.	Refinish or replace drums in axle sets.
Disc	Incorrect pad or pads loose on plate.	Replace with correct pads and plates in axle sets.
	Grease on pads.	Repair grease seal; if necessary, replace pads and plates in axle sets.
	Brake fluid on pads.	Repair caliper; if necessary, replace pads and plates in axle sets.

	Loose caliper or caliper mounting bracket.	Tighten.
Drum or disc	Rough or corroded master-cylinder bore.	Repair or replace master cylinder.
	Bind in pedal linkage.	Free linkage and lubricate.
	Faulty power-brake vacuum booster.	Replace power section (See Table 6-2.)

PULLS TO ONE SIDE

Brake Type	Probable Cause	Corrective Action
Drum	Incorrect or distorted shoes.	Replace with correct shoes and linings in axle sets.
	Incorrect lining or lining loose on shoe.	Replace with correct shoes and linings in axle sets.
	Grease on lining.	Repair grease seal; if necessary, replace shoes and linings in axle sets.
	Brake fluid on lining.	Repair wheel cylinder; if necessary, replace shoes and linings in axle sets.
	Shoe not centered in drum.	Adjust anchor pin.
	Loose or distorted backing plate.	Tighten or replace backing plate.
	Scored, hard-spotted, or out-of-round drum.	Refinish or replace drums in axle sets.
	Different sized drums on same axle.	Refinish or replace drums in axle sets.
	Water on lining.	Apply brakes a few times to dry lining.

	Sticking wheel-cylinder piston.	Repair or replace wheel cylinder.
Disc	Incorrect or loose pad.	Replace with correct pads and plates in axle sets.
	Grease on pad(s).	Repair grease seal; if necessary, replace pads and plates in axle sets.
	Brake fluid on pad.	Repair caliper; if necessary, replace pads and plates in axle sets.
	Caliper loose on bracket.	Tighten.
	Caliper piston sticking.	Repair or replace caliper.
Drum or disc	Faulty suspension.	Repair suspension system.

THUMPING AS CAR NEARS A STOP

Brake Type	Probable Cause	Corrective Action
Drum	*Out-of-round drum.*	Refinish or replace drums in axle sets.
	Bent rear axle.	Replace axle.
Disc	Excessive variation in rotor thickness.	Refinish or replace rotor.
	Excessive lateral runout in rotor.	Retorque lug bolts; if problem persists, refinish or replace rotor.
Drum or disc	Worn or damaged front-wheel lining.	Replace lining.

Brake Type	Probable Cause	Corrective Action
Drum	Bent, broken, or incorrect shoe.	Replace with correct shoes and linings in axle sets.
	Worn-out linings (shoe rubbing drum).	Replace shoes and linings in axle sets; if necessary, refinish drums.
	Foreign material imbedded in lining.	Replace shoes and linings in axle sets.
	Broken return spring, shoe hold-down pin, or hold-down spring.	Replace defective parts.
	Rough, grooved, or dry shoe ledge or backing-plate pad.	Smooth ledge and pad and lubricate with high-temperature grease.
	Cracked or threaded (lathe-marked) drum.	Replace drums in axle sets.
Disc	Damaged or incorrect pad and plate.	Replace with correct pads and plates in axle sets.
	Worn-out pads (plates rubbing rotor).	Replace pads and plates in axle sets.
	Foreign material imbedded in pad.	Replace pads and plates in axle sets.

INSUFFICIENT BRAKE-PEDAL TRAVEL

Brake Type	Probable Cause	Corrective Action
Drum	Weak or broken shoe-return spring.	Replace all return springs.
	Sticking wheel-cylinder piston.	Repair or replace wheel cylinder.

Brake Type	Probable Cause	Corrective Action
Drum or disc	Plugged master-cylinder compensating port.	Repair or replace master cylinder.
	Swollen master-cylinder cups.	Repair master cylinder, drain and flush system, refill with new DOT-3 fluid, and bleed system.

ONE BRAKE DRAGS

Brake Type	Probable Cause	Corrective Action
Drum	Weak or broken return spring.	Replace spring.
	Improper shoe adjustment.	Adjust shoes and, if necessary, repair automatic adjuster'.
	Sticking wheel-cylinder piston.	Repair or replace wheel cylinder.
	Swollen wheel-cylinder cups.	Repair or replace wheel cylinder; drain and flush hydraulic system, refill with DOT-3 fluid, and bleed system.
	Broken or distorted shoes.	Replace shoes and linings in axle sets.
Disc	Sticking caliper piston.	Repair or replace caliper.
	Swollen caliper-piston seal.	Repair caliper; drain and flush system, and refill with proper fluid. Bleed system.
Drum or disc	Loose or worn front-wheel bearings.	Tighten spindle nut; if necessary, replace bearings.

| | Defective hose or tube (preventing fluid return from wheel cylinder or caliper). | Replace defective hose or tube. |

BOTH REAR BRAKES DRAG

Brake Type	Probable Cause	Corrective Action
Drum	Frozen parking-brake cable.	Free up and lubricate cable. Replace as needed.
	Improper shoe adjustment.	Adjust shoes and, if necessary, repair automatic adjuster.

ALL BRAKES DRAG

Brake Type	Probable Cause	Corrective Action
Drum or disc	Binding brake pedal.	Free up and lubricate pedal and pivot pin.
	Soft or swollen rubber parts (due to incorrect or contaminated fluid).	Replace all rubber parts; drain and flush system and refill with proper fluid; bleed system.
	Plugged master-cylinder compensating ports.	Repair or replace master cylinder.

STOPLIGHTS DON'T COME ON

Stoplights	Probable Cause	Corrective Action
	Fuse burned out.	Replace fuse.
	Bulb burned out.	Replace bulb.
	Stoplight switch defective.	Replace switch.
	Wire loose or broken.	Tighten connectors or, if necessary, replace wire.

Switch out of alignment.	Reposition switch to allow proper brake-pedal-shaft contact.

PARKING-BRAKE INDICATOR DOESN'T WORK

Probable Cause	Corrective Action
Fuse burned out.	Replace fuse.
Bulb burned out.	Replace bulb.
Switch defective.	Replace switch.
Wire loose or broken.	Tighten connectors; if necessary, replace wire.

BRAKE-WARNING LIGHT STAYS ON

Probable Cause	Corrective Action
Failure in hydraulic system (front or rear).	Repair system; reset brake-warning light switch[2].

[2] GMC brake-warning light switch resets automatically.

BRAKE-WARNING LIGHT WON'T WORK

Probable Cause	Corrective Action
Fuse burned out.	Replace fuse.
Bulb burned out.	Replace bulb.
Switch defective.	Replace switch.
Wire loose or broken.	Tighten connectors; if necessary, replace wire.

PARKING BRAKE DOESN'T WORK

If your brake system functions properly, but your parking brake doesn't hold, either the cable is broken or it needs adjustment. If either the cable or clevis pin is defective, replace it. If cable adjustment is needed, refer to Chapter 13.

Before We Begin 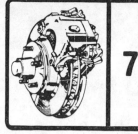 7

There are a number of cautions about brake work—both for your own safety and for the sake of the job. It would not be very helpful to save a hundred dollars on a brake-system overhaul and spend six months in the hospital recovering from the effects of a collapsed bumper jack. Nor would it be very helpful to overhaul your brakes and have them go out a week or so later.

JACKS AND JACK STANDS

Bumper jacks are designed to provide an inexpensive means of allowing us to change a tire under emergency conditions. The typical bumper jack is used perhaps once or twice during the life of the car. Bumper jacks are not intended to be mechanics' tools. The jacks have a relatively short lifespan, do not provide a stable platform, and (in many cases) have a restricted range. If at all possible, you should buy, rent, or borrow a wheeled hydraulic jack of the type favored by service stations. These jacks adjust themselves as the car is lifted. An air jack is a good choice, if you have a source of compressed air to operate it.

Observe these cautions:

- Park the car on level concrete.
- Block the offside wheels fore and aft (Fig. 7-1).
- Check the jack alignment as the car is raised. A jack may be secure at 6 in. of lift but totter precariously at 12 in.
- Never trust work under any jack with your life, regardless of its sophistication, cost, or built-in safety features. Bumper jacks are the worst of the lot, but no jack is safe. Support the weight of the car with one or more jack stands. Now, the question arises, "Where does one place a jack stand?" And the general answer is, "Under the rear axle, as near as possible to the wheel (for rear-wheel service), and under the front-suspension bracket or steering bracket (for front-wheel service)." To aid you in identifying these

Fig. 7-1. How to use a bumper jack. Note: Chocks fore and aft of rear wheel.

lift points, examine the diagrams which show these points on many late-model American-built cars: Figs. 7-2 (American Motors), 7-3 (Ford), 7-4 (Chrysler), and 7-5 (General Motors).

● After the jack stand is placed, try to shake the car by hand. Don't get under any car you can move.

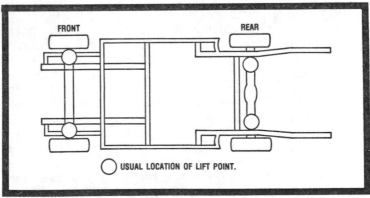

Fig. 7-2. Lift points for most American Motors cars and all jeeps.

DOING THE JOB RIGHT

Brake maintenance must never be done halfway.

You should never begin a brake-repair job unless you have every intention of doing a thorough and proper job. For example, if you replace the brake shoes (or pads), it is imperative that you overhaul the wheel cylinders. And, if you bleed the hydraulic system—a necessity when overhauling wheel cylinders—then it becomes essential that you overhaul the master cylinder. *Any adjustment or change in the brake system that alters the length of wheel-cylinder piston travel causes the piston to travel over pitted, rusted, and sedimented surfaces, which results in damage to the cups (seals). And damaged cups inevitably leak*—if not immediately, surely within a few days. The same holds true for the master cylinder; that is, bleeding the system will increase master-cylinder piston travel, and fluid leakage is the assured result.

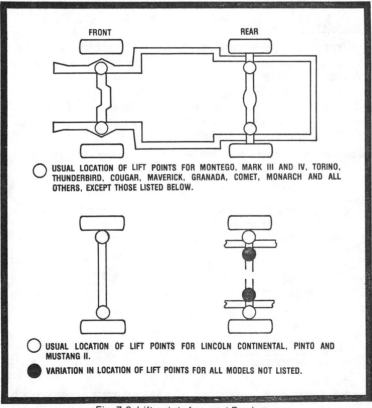

Fig. 7-3. Lift points for most Ford cars.

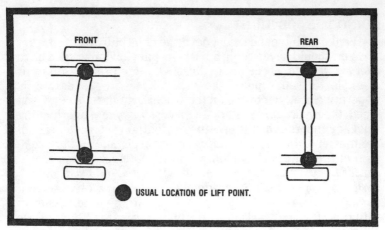

Fig. 7-4. Lift points for most Chrysler cars.

And remember, most critical brake failures are the direct result of fluid leaks.

TOOLS AND SUPPLIES

Brake work requires surprisingly little in the way of tools. We will describe the specific tool requirements for each job in appropriate places in this book. Figure 7-6 illustrates the basic tool kit that is complete enough to handle most jobs. From the left, these tools are: a socket driver (you may prefer a ratchet handle to the breakover bar shown); sockets open-end and box-end wrenches (combinations with half-inch on one end,

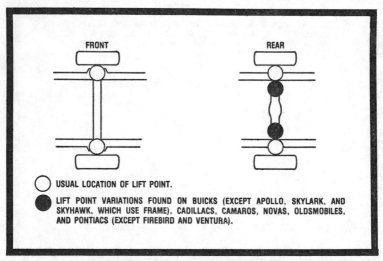

Fig. 7-5. Lift points for most GM cars.

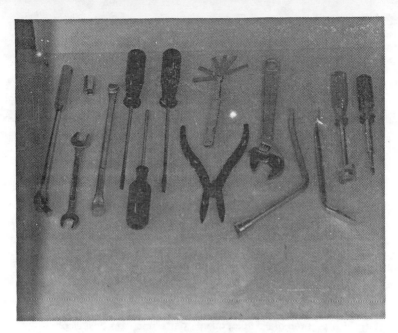

Fig. 7-6. Basic brake tools.

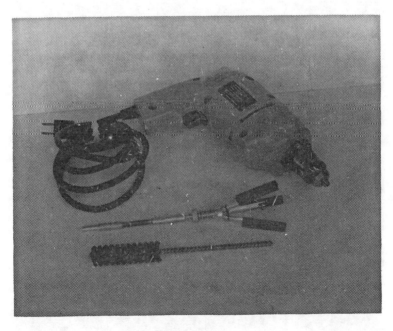

Fig. 7-7. Drill motor and hones.

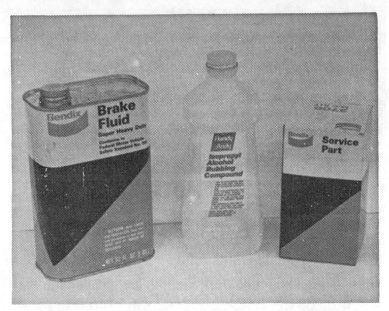

Fig. 7-8. Approved brake fluid, quality parts, and isopropyl alcohol.

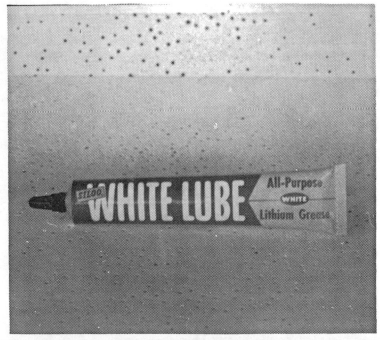

Fig. 7-9. Lithium grease—a good general-purpose lubricant.

nine-sixteenths on the other will fit most brake fasteners); screwdrivers (flat and Phillips bits); a feeler gage; snapring pliers; an adjustable wrench; a brake-spring tool (the one shown is designed for Bendix springs but will work on any that are mounted to a pin); an adjusting tool (in a pinch you can use a large screwdriver); a vacuum-check valve socket; and brake-shoe retainer spring tool (not really necessary but speeds the work).

For honing the master and wheel cylinders, you will need a ¼ in. drill motor and a hone. Two types are shown in Fig. 7-7: the adjustable type and the newer *fixed* type. In addition you should have a bottle brush to clean the cylinder bores after honing.

You will also need a quart of high-quality brake fluid, wood alcohol (for cleaning) and, of course, whatever parts are required (Fig. 7-8). Purchase name-brand parts since there is little regulation of replacement parts quality. The best guarantee you have is the reputation of the maker. White grease (Fig. 7-9) is required to lubricate parking-brake mechanisms and other purely mechanical assemblies. And you should purchase several cotter pins for the front-wheel spindles (Fig. 7-10). Using the same cotter pin twice is always unwise, especially on something as critical as the spindle nut.

Fig. 7-10. Always use a new cotter pin on the spindle nuts.

Step-by-Step Master Cylinder and Booster Overhaul

8

Two types of master cylinders are found on American cars—the now-obsolete single-system cylinder and the dual-system cylinder.

SINGLE-SYSTEM REPAIRS

Before beginning actual work, make sure you have all the necessary tools and materials at hand to do the job. Then read through the sequence once to become familiar with the operations required.

Parts and Materials

- Master-cylinder repair kit (for your specific car).
- Quart of DOT-3 brake fluid.
- Enough isopropyl alcohol or brake-parts cleaning fluid—to bathe all master-cylinder parts that are to be reused, and to flush the system if necessary. Caution: Don't use gasoline, kerosene, distillate, or any other petroleum product in or on any part of a brake system—these materials contaminate and will distort the cups and seals.
- Empty container in which to drain waste fluids.

Tools

- Open-end or tube-fitting wrench, properly sized to remove the tubing fittings from your master-cylinder housing.
- Open-end, box-end, or socket wrench, properly sized to remove the bolts or nuts that attach your master cylinder to the firewall, power-brake assembly, or mounting bracket.

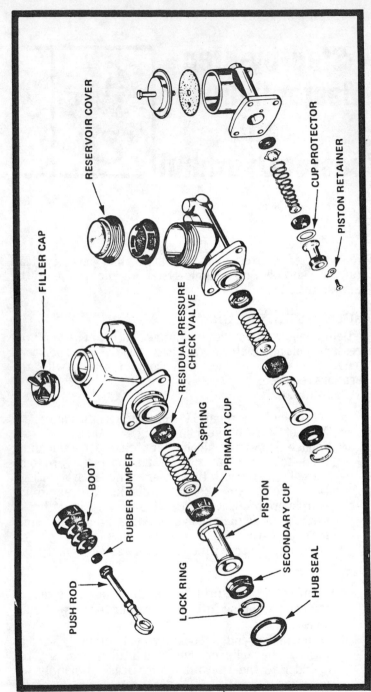

Fig. 8-1. Exploded view of three common types of single-system master cylinders.

RESERVOIR COVER

FILLER CAP

CUP PROTECTOR

PISTON RETAINER

RESIDUAL PRESSURE CHECK VALVE

BOOT

RUBBER BUMPER

PUSH ROD

SPRING

PRIMARY CUP

PISTON

SECONDARY CUP

LOCK RING

HUB SEAL

116

- Small screwdriver or a snapring pliers to remove snapring (not used in all single-system master cylinders).
- Brake-cylinder hone (or crocus cloth).
- Electric drill to operate hone.
- Narrow-blade or wire feeler gage (0.006 in.).

Disassembly

Three single-system cylinders are shown in exploded view in Fig. 8-1. Note that not all steps are appropriate to all cylinders, so skip any steps not applicable to your specific system.

Step 1. Use open-end or tube-fitting wrench to remove hydraulic-line fittings from master cylinder and allow fluid from cylinder to drain into can or jar for later disposal.

Step 2. If stoplight switch is installed on master cylinder, remove it by detaching wires and turning switch counterclockwise to unscrew it from cylinder housing.

Step 3. Use open-end, box-end, or socket wrench to remove all bolts or nuts that attach master cylinder to power brake or firewall.

Step 4. Lift out master cylinder assembly. *Note*: On some models, the cylinder-activating rod must be detached from brake pedal (inside passenger compartment) and removed with master-cylinder assembly. Remove pushrod from piston as shown in Fig. 8-2.

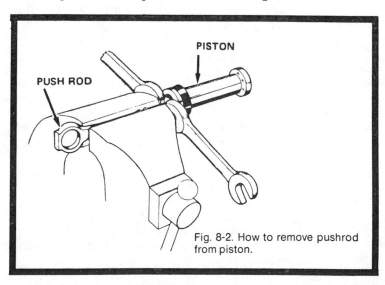

PISTON

PUSH ROD

Fig. 8-2. How to remove pushrod from piston.

Step 5. Place master-cylinder assembly in normal position on bench; remove filler cap (or reservoir cover); pour all fluid from reservoir. *Note*: Do not reuse fluid.

Step 6. Remove snapring from inside cylinder bore (at open or rear end).

Step 7. Remove all parts from inside cylinder bore and lay them out on bench in the exact order of their removal.

Step 8. The last part removed from the cylinder bore in step 7 should be (on most types) a rubber ring that seats at the bottom of the bore; it is the residual-pressure valve seat.

Step 9. Clean the master cylinder thoroughly inside and out with alcohol or brake-parts cleaning fluid. *Do not use any kind of petroleum product*. Water is a satisfactory cleaning fluid *provided cylinder is allowed to dry thoroughly before it is reassembled*.

Step 10. Lubricate hone and cylinder bore with brake fluid. Attach hone to electric drill. Hone cylinder to remove all rust and pits. *Note*: Don't hone any more than is absolutely necessary.

Step 11. Clean the master cylinder again, as in step 9.

Step 12. Lubricate the *old* piston with new brake fluid; insert into cylinder bore; attempt to push 0.006 in. feeler gage between piston and cylinder wall; if you cannot, proceed with reassembly of master cylinder; if you *can* insert feeler gage between piston and cylinder wall, the master cylinder is too worn to work satisfactorily and must be discarded—throw it away, return your repair kit, and get a new master-cylinder assembly.

Reassembly

Step 13. Be sure that all parts to be assembled (new as well as old) are thoroughly clean. (Contamination is a hydraulic system's worst enemy.)

Step 14. Lubricate all parts to be assembled into cylinder bore with new brake fluid; also lubricate cylinder bore.

Step 15. Place new residual-pressure check valve seat into bottom of bore (some models).

Step 16. Place new residual-pressure check valve on end of spring and drop this assembly into bore, valve first.

Step 17. Carefully place the primary cup inside cylinder, cup lip first; press cup into cylinder about ½ in.—use finger pressure only.

Step 18. Drop cup protector (a thin metal washer) into cylinder. *Note*: Cup protectors are not provided for some models.

Step 19. Install piston assembly; insert end with bleeder holes first, which means end with attached rubber cup goes last; install carefully—do not force; primary cup must not be damaged.

Step 20. Install piston retainer and screw or snapring (whichever your assembly has) and stoplight switch.

Step 21. Install master cylinder into car; be sure all bolts and fittings are snugg and that the activating rod is attached.

Step 22. Refill system with brake fluid in accordance with instructions in Chapter 12.

Step 23. Test your brake system carefully, as prescribed in Chapter 15.

DUAL-SYSTEM REPAIRS

Before beginning actual work, make sure you have all the necessary tools and materials required to do the job. Read through the disassembly and reassembly instructions once to become familiar with the procedures involved.

Parts and Materials

- Master-cylinder repair kit (for your specific car).

- Quart of DOT-3 brake fluid.
- Enough isopropyl alcohol or brake-parts cleaning fluid to bathe all master cylinder parts that are to be reused, and to flush the system if necessary. *Caution*: don't use gasoline, kerosene, distillate, or any other petroleum product in or on any part of a brake system—these materials contaminate and will distort the cups and seals.

Tools

- Open-end or tube-fitting wrench, properly sized to remove the tubing fittings from your master-cylinder housing.
- Open-end, box-end, or socket wrench, properly sized to remove the bolts (or nuts) that attach your master cylinder to mounting bracket, firewall, or power-brake assembly.
- Narrow-blade screwdriver (approx. 3/16 in.).
- Adjustable wrench with a 1 in. bite.
- Brake-cylinder hone or crocus cloth.

Fig. 8-3. Bendix dual-system master cylinder (early model).

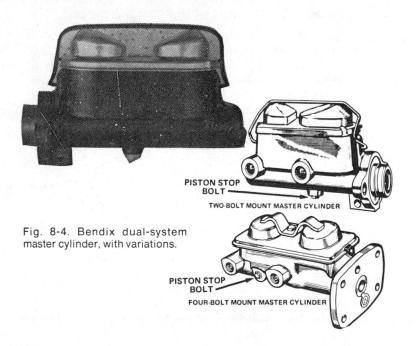

PISTON STOP
BOLT

TWO-BOLT MOUNT MASTER CYLINDER

Fig. 8-4. Bendix dual-system
master cylinder, with variations.

PISTON STOP
BOLT

FOUR-BOLT MOUNT MASTER CYLINDER

- Electric drill to operate hone.
- Narrow-blade or wire feeler gage (0.006 in.).

Refer to Figs. 8-3, 8-4, and 8-5; select the illustration most nearly resembling your master cylinder in outward appearance—it will most nearly resemble it in inward appearance as well.

Master-Cylinder Disassembly Early Bendix Dual-System

Step 1. Use tube-fitting wrench or an open-end wrench to disconnect hydraulic lines from master cylinder. Do not bend steel hydraulic lines more than absolutely necessary. A small amount of fluid may drain from master cylinder and lines—use cloth to absorb this fluid and to mop it off other car parts.

Step 2. Use snug-fitting wrench to remove all bolts or nuts that secure master cylinder to power brake firewall, or mounting bracket.

Step 3. Lift out master cylinder, and place it in an upright position on bench.

Step 4. Remove cover(s) (Fig. 8-6) and set aside; pour all fluid out of reservoirs into a container for later disposal. Do not reuse old fluid.

Step 5. If master cylinder includes a piston-stop screw, remove it and set it aside.

Step 6. Remove side outlet plug and, from beneath it, the seal, spring, and residual-pressure check valve.

Step 7. Use snapring pliers (or small screwdriver) to remove snapring—sometimes called a lockring—from cylinder bore. Not all models have this ring.

Fig. 8-5. Delco-Moraine dual-system master cylinder (early model).

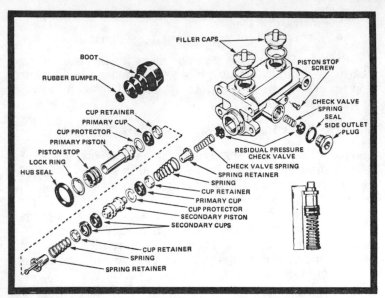

Fig. 8-6. Exploded view of Bendix dual-system master cylinder (early model).

Step 8. Remove and disassemble piston parts from cylinder bore; lay them out on bench in proper order, as shown in the exploded view.

Step 9. Thoroughly clean master-cylinder housing, inside and out, and all metal parts to be reused, with denatured (isopropyl) alcohol or brake-parts cleaning fluid. Do not use any kind of petroleum product. Clean water is a satisfactory cleaning fluid provided cylinder is allowed to dry thoroughly before reassembly.

Step 10. Attach brake-cylinder hone to electric drill; lubricate hone and cylinder bore liberally with new brake fluid; hone cylinder to remove rust, pits, and burrs. Do not hone any more than is absolutely necessary.

Step 11. Clean master-cylinder housing as in Step 9.

Step 12. Lubricate the old piston with new brake fluid, then insert it into cylinder bore; attempt to push a .006 in. feeler gage between piston and cylinder wall. If you cannot, remove piston and proceed to Step 13. If you can, cylinder is too worn to function properly and must be discarded—throw it away, return your repair kit and get a new master-cylinder assembly. Then proceed to Step 26.

Master Cylinder Reassembly

Step 13. Be sure all parts to be assembled (new as well as old) are thoroughly cleaned.

Step 14. Assemble new parts on secondary piston, as shown in Fig. 8-6. Insert screw into piston until piston length is equal to gage height—gage is included in master-cylinder parts kit.

Step 15. Lubricate new piston stop with new brake fluid, and assemble it to primary piston:

(a) If stop is made of plastic, slide it on piston, either end first.

(b) If stop is made of brass (Cadillac) slide it on piston, large bore first.

(c) If stop is made of brass (American Motors), slide it on piston, small bore first.

Step 16. Assemble other new parts on primary piston, as shown in the exploded view.

Step 17. Lubricate cylinder bore with new brake fluid.

Step 18. Dip secondary-piston assembly in new brake fluid, then assemble return spring, spring retainer, check-valve spring, and residual-check valve on end of secondary piston, as shown.

Step 19. Lubricate inside of piston installation tube (provided in parts kit) with new brake fluid; slide secondary-piston assembly out of tube and into cylinder bore, as shown. Insert tube into cylinder bore, check valve first; then use finger or screwdriver to slide assembly out of tube into cylinder bore.

Step 20. Dip primary-piston assembly in new brake fluid; insert this assembly into cylinder bore.

Step 21. Press in on primary piston; insert snapring into groove in cylinder bore (if your master cylinder has one).

Step 22. Press in on primary piston; insert and tighten piston-stop screw. Release piston—it should move freely for approximately 1½ in.

Step 23. Assemble new check-valve spring, residual-pressure check valve, and seal beneath old side outlet plug. Install and tighten plug.

Step 24. If your master cylinder has a hub seal (as shown in Fig. 8-6), install new one into groove on exterior of hub.

Step 25. If your master cylinder uses a rubber pushrod bumper, install new one inside primary piston bore.

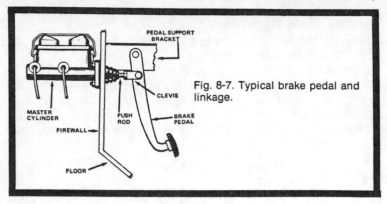

Fig. 8-7. Typical brake pedal and linkage.

Step 26. If your car does not have power brakes, remove old dustboot from pushrod and install new one.

Step 27. Bench bleed master cylinder. (Refer to Chapter 12 for bench-bleeding instructions.)

Step 28. Install master cylinder on car; attach hydraulic lines and tighten them firmly.

Step 29. Refill master cylinder and bleed system. (Refer to Chapter 12 for instructions.)

Step 30. Test brakes. (Refer to Chapter 15 for instructions).

Late-Model Bendix Dual System
Master-Cylinder Disassembly

Step 1. If your car does not have power brakes, disconnect master-cylinder pushrod from brake pedal (remove clevis pin from inside passenger compartment, as shown in Fig. 8-7).

Step 2. Disconnect hydraulic lines from master cylinder. Mop up spilled brake fluid.

Step 3. Remove bolts or nuts that secure master cylinder to firewall mounting bracket, or power brake.

Step 4. Lift out master cylinder, place it on bench in upright position.

Step 5. Remove cover and diaphragm and pour old brake fluid out.

Step 6. Remove piston-stop bolt (Fig. 8-8).

Step 7. Press in on pushrod and use snapring pliers or small screwdriver to remove snapring from cylinder bore or piston retainer from mounting flange, whichever your master cylinder uses.

Step 8. Remove all piston parts from cylinder bore—they should come out in the order shown in Fig. 8-8. Keep them in this order at all times.

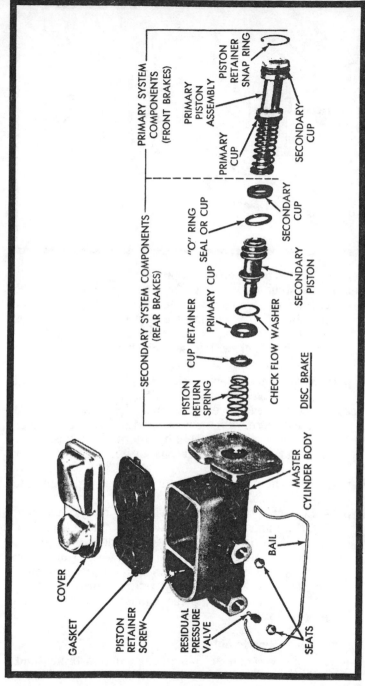

Fig. 8-8. Bendix dual-system master cylinder (late model). (Courtesy Chrysler Corp.

125

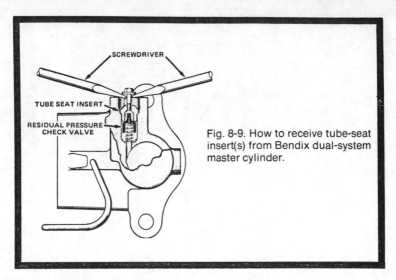

Fig. 8-9. How to receive tube-seat insert(s) from Bendix dual-system master cylinder.

Step 9. Disassemble secondary piston.

Step 10. If either outlet part contains a residual-pressure check valve, remove tube seat from that part. (Refer to Table 2-1 to determine whether or not check valve is present.) To remove tube seat, thread a No. 6 ×¾ sheet-metal screw, into seat as shown in Fig. 8-9 and pry out with screwdrivers.

Step 11. Thoroughly clean master-cylinder housing and all metal parts to be reused.

Step 12. Attach hone to electric drill; lubricate hone and cylinder bore with new brake fluid. Hone cylinder bore. *Note*: never hone any more than absolutely necessary to remove rust, pits, or burrs from cylinder wall.

Step 13. Thoroughly reclean master cylinder, especially the cylinder bore.

Step 14. Lubricate the old piston with new brake fluid, and insert it into cylinder bore. Attempt to push a 0.006 in. feeler gage between piston and cylinder wall. If you cannot, remove piston and proceed to step 15. If you can, cylinder is too worn to function properly and must be discarded—throw it away, return your repair parts kit, and get a new master-cylinder assembly. Then proceed to step 23.

Master Cylinder Reassembly

Step 15. Dip new secondary piston parts in new brake fluid, then assemble piston as shown in Fig. 8-8.

Step 16. Lubricate cylinder bore and all piston parts with brake fluid.

Step 17. Install secondary-piston assembly in cylinder bore, spring-end first.

Step 18. Install new primary piston into cylinder bore.

Step 19. Press in on primary piston, and install snapring into cylinder bore or retainer on mounting flange, whichever is used on your master cylinder.

Step 20. Install pushrod (if your master cylinder has one) in primary piston.

Step 21. Again, press in on primary piston and insert and tighten piston stopscrew. Piston should move approx. 1 in. after stopscrew is installed.

Step 22. If tube seats were removed, insert new check-valve spring into a new check valve and drop both into outlet port, spring-end first—spring should be seated in bottom of port. Now, position tube seat in port, then press it into place by threading a tube nut into port. Repeat this step for second port if necessary. Remove tube nut.

Step 23. Bench bleed master cylinder. (Refer to Chapter 12 for instructions on bench bleeding.)

Step 24. Install master cylinder onto car; attach hydraulic lines and tighten them firmly.

Step 25. If your car does not have power brakes, attach pushrod to brake pedal.

Step 26. Refill master cylinder and bleed brakes.(Refer to Chapter 12 for instructions on brake bleeding.)

Step 27. Test brakes. (Refer to Chapter 15 for instructions on brake testing.)

Master Cylinder Disassembly
Early Delco-Moraine Dual-Systems

Step 1. Use tube-fitting or open-end wrench; disconnect hydraulic lines. Bend these lines no more than absolutely necessary.

Step 2. Use snug-fitting wrench to remove all nuts or bolts that secure master cylinder to firewall mounting bracket or power-brake assembly.

Step 3. Lift out master cylinder and place it in upright position on bench.

Step 4. Remove both filler caps (or cover) and gaskets. Pour old brake fluid out of both reservoirs.

Step 5. Remove piston stopscrew. (See Fig. 8-10 for an exploded view of this master cylinder.)

Step 6. Use snapring pliers or small screwdriver to remove snapring from flanged end of cylinder bore.

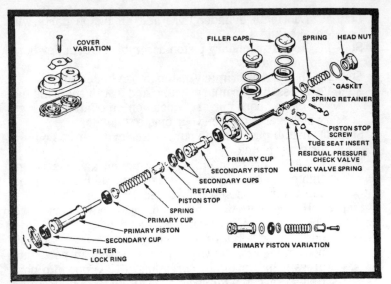

Fig. 8-10. Exploded view of Delco-Moraine dual-system master cylinder (early model).

Step 7. If your master cylinder has one, remove head nut from opposite end of cylinder housing.

Step 8. Remove all parts from cylinder head. They should come out in order shown in the drawing—keep them in this order.

Step 9. Remove tube seats from outlet ports. Thread a No. 6 × ¾ in. sheet-metal screw, into seat, as shown previously in Fig. 8-9; pry out with screwdrivers. Discard the tube-seat assemblies.

Step 10. Disassemble both the primary and the secondary pistons. Keep parts in proper order, but substitute new parts from kit for old parts.

Step 11. Thoroughly clean master-cylinder housing and all metal parts that are to be reused.

Step 12. Attach hone to electric drill; lubricate hone and cylinder bore with new brake fluid.

Step 13. Hone cylinder bore to remove rust, burrs, and pits, but do not hone bore any more than is absolutely necessary.

Step 14. Reclean master-cylinder housing, especially cylinder bore.

Step 15. Lubricate old piston and cylinder bore with new brake fluid. Drop piston into cylinder bore.

Step 16. Attempt to insert a 0.006 in. feeler gage between piston and cylinder wall. If you cannot, remove piston and proceed with step 17. If you can, cylinder is too worn to function properly and must be discarded—throw it away, return your repair parts kit, and get a new master-cylinder assembly. Then proceed to step 17.

Master Cylinder Reassembly

Step 17. Insert two new check-valve springs into the two new check valves; drop them into the two outlet ports, spring-ends first. Springs must be seated in ports.

Step 18. Position tube seats in ports, then press each into place by threading a tube nut into ports, one at a time. Remove tube nut.

Step 19. Assemble new parts on secondary piston. *Note*: Two aft secondary cups are assembled back to back.

Step 20. Assemble new parts on primary piston.

Step 21. Lubricate cylinder bore with new brake fluid; dip secondary piston in new brake fluid; insert piston into cylinder bore, primary-cup end first. *Note*: If your master cylinder does not have a head nut, then the secondary piston spring and spring retainer must be installed into cylinder bore before secondary piston is inserted.

Step 22. Push secondary piston into cylinder until piston end is even with first head-nut thread; then insert and tighten the piston stopscrew.

Step 23. Place spring retainer on end of spring and insert this assembly, spring first, into head-nut well. Install head nut and tighten firmly—to 105 ft-lb of torque.

Step 24. Dip primary-piston assembly in new brake fluid and insert into cylinder bore, spring-end first.

Step 25. Install snapring into cylinder bore.

Step 26. Install new filter on hub.

Step 27. Bench bleed master cylinder. (Refer to Chapter 12 for instructions on bench bleeding.)

Step 28. Install master cylinder on car. Be sure hydraulic lines are tightly fitted.

Step 29. If you did not use torque wrench on head nut, retighten it.

Step 30. Refill master-cylinder reservoir and bleed it. (Refer to Chapter 12 for instructions.)

Step 31. Test brakes. (Refer to Chapter 15 for instructions on brake testing.)

Delco-Moraine Dual Systems
Master Cylinder Disassembly

Step 1. If your car does not have power brakes, disconnect master-cylinder pushrod from brake pedal (remove

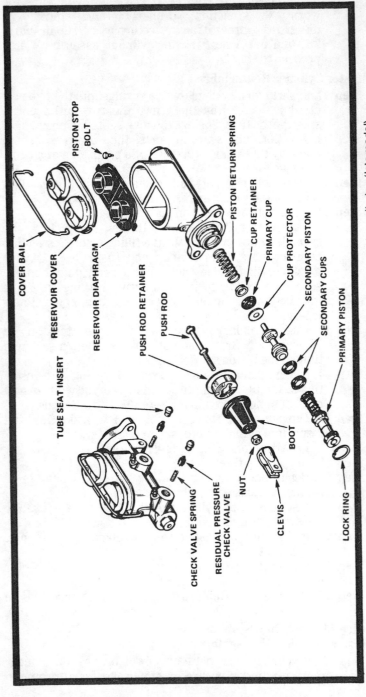

COVER BAIL

PISTON STOP BOLT

RESERVOIR COVER

RESERVOIR DIAPHRAGM

PISTON RETURN SPRING

CUP RETAINER

PRIMARY CUP

CUP PROTECTOR

SECONDARY PISTON

SECONDARY CUPS

PRIMARY PISTON

LOCK RING

TUBE SEAT INSERT

PUSH ROD RETAINER

PUSH ROD

BOOT

NUT

CLEVIS

CHECK VALVE SPRING

RESIDUAL PRESSURE CHECK VALVE

Fig. 8-11. Exploded view of Delco-Moraine dual-system master cylinder (late model).

clevis pin from inside passenger compartment, as shown back in Fig. 8-7.)

Step 2. Disconnect hydraulic lines from master cylinder. Mop up spilled brake fluid.

Step 3. Remove bolts or nuts that secure master cylinder to firewall, mounting bracket, or power-brake assembly.

Step 4. Lift out master cylinder, place it on bench in upright position.

Step 5. Remove cover and diaphragm, pour old brake fluid out and dispose of it.

Step 6. Remove piston stopscrew. (See Fig. 8-11 for exploded view of this master cylinder.)

Step 7. Press in on pushrod and use snapring pliers or small screwdriver to remove snapring from cylinder bore or piston retainer from mounting flange, which ever your master cylinder uses.

Step 8. Remove all piston parts from cylinder bore—they should come out in the order shown in the drawing. Keep them in this order at all times.

Step 9. Disassemble secondary piston.

Step 10. If either outlet port contains a residual-pressure check valve, remove tube seat from that part. (Refer to Table 2-1 to determine whether or not check valve is present.) To remove tube seat, thread a No. 6 × ¾ in. sheet-metal screw seat as shown in Fig. 8-9. Pry out with screwdrivers.

Step 11. Thoroughly clean master-cylinder housing and all metal parts to be reused.

Step 12. Attach hone to electric drill; lubricate hone and cylinder bore with new brake fluid. Hone cylinder bore. *Note*: Never hone any more than is absolutely necessary to remove rust, pits, or burrs from cylinder wall.

Step 13. Thoroughly reclean master cylinder, especially the cylinder bore.

Step 14. Lubricate the old piston with new brake fluid, then insert it into cylinder bore. Attempt to push your 0.006 in. feeler gage between piston and cylinder wall; if you cannot, remove piston and proceed to step 15. If you can, cylinder is tto worn to function properly and must be discarded—throw it away, return your repair parts kit, and get a new master-cylinder assembly. Then proceed to step 23.

Master Cylinder Reassembly

Step 15. Dip new secondary piston parts in new brake fluid, then assemble piston, as shown in Fig. 8-11.

Step 16. Lubricate cylinder bore and all piston parts with brake fluid.

Step 17. Install secondary piston assembly in cylinder bore, spring-end first.

Step 18. Install new primary piston into cylinder bore.

Step 19. Press in on primary piston, and install snapring into cylinder bore or retainer on mounting flange, whichever is used on your master cylinder.

Step 20. Install pushrod (if your master cylinder has one) into primary piston. If pushrod is like variation shown in Fig. 8-11, install new boot on rod, then place new retainer on rod end; press rod into primary piston until you feel retainer snap into place.

Step 21. Press in on primary piston and insert and tighten piston stopscrew. Piston should move about 1 in. after stopscrew is installed.

Step 22. If tube seat was removed, insert new check-valve spring into a new check valve and drop both into outlet port, spring-end first—spring should be seated in bottom of port. Now, position tube seat in port, then press it into place by threading a tube nut into port. Repeat this operation for second port if necessary. Remove tube nut.

Step 23. Bench bleed master cylinder (Refer to Chapter 12 for instructions on bench bleeding.)

Step 24. Install master cylinder on car; attach hydraulic lines and tighten them firmly.

Step 25. If your car does not have power brakes, attach pushrod to brake pedal.

Step 26. Refill master cylinder and bleed brakes. (Refer to Chapter 12 for instructions on brake bleeding.)

Step 27. Test brakes. (Refer to Chapter 15 for instructions on brake testing.)

VACUUM-BOOSTER REPAIR

As suggested earlier, there are two types of failure one might encounter in power brakes: (1) a defective check valve or (2) a leaking power section. In either case, replacement is the answer.

Check Valve Replacement

Step 1. Remove vacuum hose from vacuum-check valve. (See Fig. 8-12.)

Step 2. Remove and discard check valve from power section; valve screws out counterclockwise.

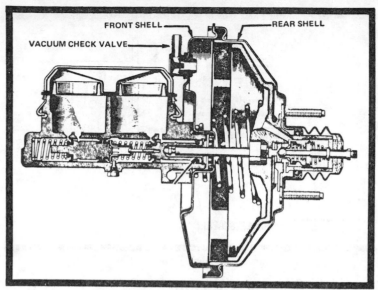

Fig. 8-12. Typical vacuum-check valve.

Step 3. Use new gasket; install and tighten new vacuum check valve.

Step 4. Attach vacuum hose. Use new hose and clamps if necessary.

Power Section Replacement

Step 1. Disconnect vacuum hose from vacuum check valve. (See Fig. 8-13.)

Step 2. Remove master cylinder from power section. It is normally mounted to power section by two or four stud bolts; use open-end or box-end wrench to remove nuts from these bolts. *NOTE:* Unfortunately, many car manufacturers hang other paraphernalia from the power-section mounting bracket, which is normally attached directly to the firewall by studs. Be sure everything goes back on these bolts in the same order you take it off.

Step 3. If necessary, disconnect pushrod from brake pedal. Remove pin from clevis, as shown in Fig. 8-13.

Step 4. Install new power section. If new section does not include pushrod, install old one removed with old section.

Step 5. Install master cylinder on power section.

Step 6. Attach vacuum hose to vacuum check valve.

Step 7. Test brakes. (Refer to Chapter 15 for instructions.)

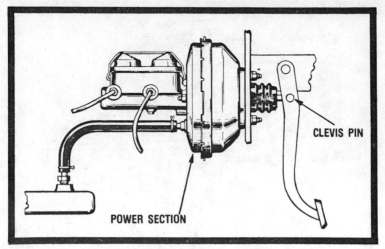

Fig. 8-13. Typical power section.

PEDAL AND LINKAGE REPAIR

There is not much that can be done for or to your brake pedal and linkage. Thus, the only "tool" required is your common sense: when the pad on the pedal looks worn, replace it. And, if the pedal feels wobbly or stiff inspect the hinge and clevis pin—but these ride on graphite-impregnated bushings and almost never need attention.

Step-by-Step Drum Brake Overhaul

9

If you are not yet an old hand at drum-brake repair, remember the following ground rules: (1) Always overhaul brakes in *axle sets*; that is, if you do one rear wheel, do both rear wheels—and if you do one front wheel, do both front wheels; (2) do one wheel at a time (disassemble, repair, and reassemble one wheel of an axle set before you touch the other—the other will always be there as a pattern, and a pattern can be a useful thing to have at hand, even for the experienced).

TOOLS AND SUPPLIES

The tool requirements vary a bit between front and rear drums. The following list includes only those items needed for overhaul of rear-wheel drum brakes:

- Wheel chocks (2).
- Jack.
- Jack stand.
- Lug wrench (bumper-jack handle usually doubles in this role).
- Brake-spring tool.
- Medium- or broad-blade screwdriver.
- Wheel-cylinder hone (or crocus cloth).
- Electric drill to operate hone.
- Tube or can of white brake lubricant.
- Replacement shoe-and-lining set (one axle set).
- Wheel-cylinder repair parts kit (one axle set). In addition, you will need all the items listed in Chapter 15 for bleeding your brake system. You should also overhaul your master cylinder any time wheel

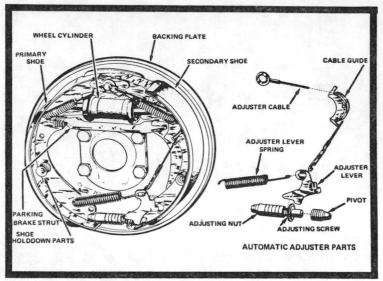

Fig. 9-1. Bendix duoservo brake with automatic adjust.

cylinders are overhauled, so refer back to Chapter 8 for items needed for master-cylinder overhaul.

No lug wrench is required for front-wheel drum-brake overhaul, but the following items are required in addition to those listed above:

- Pliers (preferably needlenose pliers).
- Adjustable wrench with large enough bite to fit wheel nut.
- Can of wheel-bearing grease.

HOW YOU DO IT

The eight basic drum-brake designs used on American cars since World War II are illustrated in the following sections. Select the one identical to yours, then refer to the appropriate set of instructions.

Bendix Duoservo Brake

There are two types of Bendix automatic-adjusting duoservo brakes. Examine Figs. 9-1 and 9-2 to determine which is applicable to your car. The sequence shown first is for Bendix duoservo brakes with *standard* automatic adjusters, described in Fig. 9-1. The second sequence is for Bendix duoservo brakes with *application-type* automatic adjuster (Fig. 9-2).

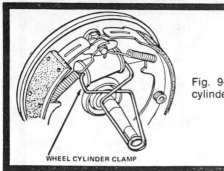

Fig. 9-2. How to use a wheel-cylinder clamp.

WHEEL CYLINDER CLAMP

Standard Automatic Adjuster Brakes. Refer to Fig. 9-1 from time to time as you perform the following steps.

Step 1. Be sure car is parked on level, hard surface.

Step 2. Place chocks fore and aft of wheel on same side of car as the one to be overhauled. When rear wheel is to be worked on do not set parking brake.

Step 3. Raise corner of car with jack.

Step 4. Place jack stand under axle or suspension. (Refer to Chapter 7 for proper jack points for your car.) NOTE: Bumper jacks are untrustworthy objects, so never depend on one. If you do not have a jack stand, place something under the frame strong enough to support the weight of the car, such as a green oak log. But do not work beneath your car while it is precariously perched atop anything so tottery as a bumper jack (or as inherently unreliable as any jack).

Step 5. Lower car weight onto jack stand; be sure wheel still clears the ground.

Step 6. Pry hubcap off wheel; place hubcap upside down nearby to serve as a receptacle for lug nuts and brake parts.

Step 7a. *Rear Wheel Only*:

(1) Loosen and remove lug nuts.

(2) Remove wheel.

(3) Remove drum from brake—it may require a tap or two to free drum from hub flange. In fact, it may be necessary to adjust brake shoes away from drum; if so, refer to Figs. 9-3 or 9-4 for visual instructions.

Step 7b. *Front Wheel Only*:

(1) Pry grease retainer from hub.

(2) Remove cotter pin.

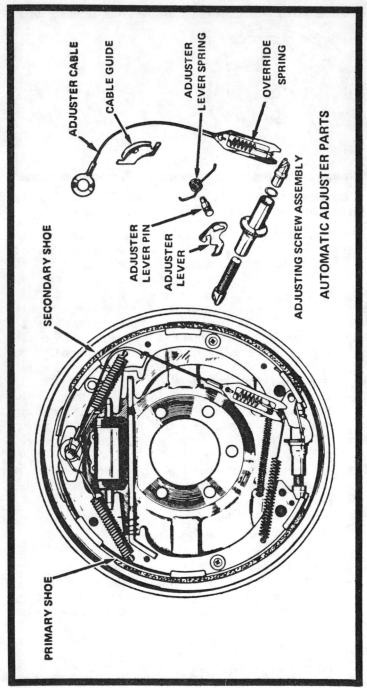

ADJUSTER CABLE

CABLE GUIDE

ADJUSTER LEVER SPRING

OVERRIDE SPRING

ADJUSTER LEVER PIN

ADJUSTER LEVER

ADJUSTING SCREW ASSEMBLY

AUTOMATIC ADJUSTER PARTS

SECONDARY SHOE

PRIMARY SHOE

Fig. 9-3. Bendix duoservo with application adjust.

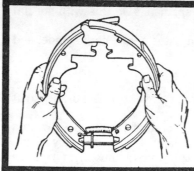

Fig. 9-4. Position for shoe removal and installation.

(3) Remove axle nut—turn it counterclockwise with an adjustable wrench while holding wheel with other hand.

(4) Remove grease retainer.

(5) Wiggle wheel and pull out on it at the same time in order to work outer bearing outward on axle (spindle). Remove bearing.

(6) Remove wheel (and outer bearing cup) from spindle.

(7) Remove inner bearing, cup, and grease retainer from spindle.

(8) Set bearings, cups, and retainers in a clean solvent or kerosene. *CAUTION*: Such solvents are both flammable and toxic; keep them away from open flames, sparks, and out of the reach of children.

(9) Wash spindle and wheel hub with solvent.

Step 8. Carefully note position of all parts of brake assembly so that they can be installed in their proper positions.

Step 9. Install a wheel-cylinder clamp (Fig. 9-5) if you have one.

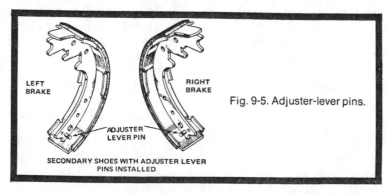

LEFT BRAKE

RIGHT BRAKE

ADJUSTER LEVER PIN

Fig. 9-5. Adjuster-lever pins.

SECONDARY SHOES WITH ADJUSTER LEVER PINS INSTALLED

Step 10. Remove shoe-return springs. (Do this with special wrench if available.)

Step 11. Remove automatic-adjuster cable, cable guide, and adjuster lever.

Step 12. Remove shoe holddown parts.

Step 13. Remove anchor plate.

Step 14. On rear wheels only, remove parking-brake strut with its antirattle spring.

Step 15. On rear wheels only, remove horseshoe clip and disconnect parking-brake lever from secondary shoe.

Step 16. Spread shoes at top to clear piston links; remove both shoes and adjusting screw from secondary shoe along with adjuster-lever spring.

Step 17. Remove anchor plate.

Step 18. Spread shoes wide apart at bottom to clear ends of shoe webs from wheel-cylinder piston pins, then slip shoes off anchor pin.

Step 19. Wash backing plate with water.

Step 20. Check shoe support points—they should be smooth and free of rough spots. Clean and sand if necessary and apply white lubricant.

Step 21. Check the adjusting-screw threads—be sure nut turns freely on screw. As necessary, clean or replace it. Apply white lubricant to threads.

Step 22. Clean all springs then check them for loss of tension: drop them from four or five feet on concrete floor or walkway. A good one lands with a thud; a bad one sounds "springy." Replace any bad ones.

Step 23. Check all brake parts; replace any that are badly worn or damaged.

Step 24. Overhaul wheel cylinder (refer to the instructions at the end of this chapter).

Step 25. On rear wheels only, assemble parking-brake lever on new secondary shoe; secure with horseshoe clip.

Step 26. Assemble secondary and primary shoe on backing plate. Hold shoes in position against anchor pin and lower web ends into slots in piston pins.

Step 27. Secure shoes with holddown parts.

Step 28. Assemble antirattle spring on parking-brake strut; position strut between parking-brake lever and primary-shoe web.

Step 29. Install anchor plate and adjuster-cable loop on anchor pin.

Step 30. If you used a wheel cylinder clamp, remove it.

Step 31. Install primary-shoe return spring.

Step 32. Position cable guide on secondary shoe web; then install secondary-shoe return spring.

Step 33. Insert adjusting screw between lower ends of shoe webs.

Step 34. Attach short-hook end of adjuster-lever spring to primary-shoe web.

Step 35. Attach other end of adjuster-lever spring and adjuster cable in hole of adjuster lever.

Step 36. Assemble adjuster lever to secondary-shoe web.

Step 37. Inspect brake drum. If necessary, have it turned or replace it. Refer to Chapter 4 for information on drum damage and repair and replacement data.

Step 38. Measure or estimate drum diameter; then adjust shoes to about ¼ in. smaller diameter so that drum will slide over easily. Adjust shoes by holding adjuster lever clear of star wheel and turning adjusting screw.

Step 39. Install drum over shoes on hub.

(a) *Rear Wheel Only*:

Replace drum on flange; then install wheel on lug bolts—tighten lug nuts.

(b) *Front Wheel Only*:

Pack wheel bearings as described at the end of this chapter. Replace retainers, bearings, and wheel on spindle in reverse order of removal (step 7). Rotate wheel slowly and hand-tighten wheel nut, then back off only as far as necessary to insert cotter pin. Flare pin ends.

Step 40. Remove jack stand and proceed to next wheel. Repeat steps 1 through 40.

Step 41. When all wheels to be overhauled are completed, overhaul master cylinder.

Step 42. Bleed system as required. (Refer to Chapter 12 for instructions.)

Step 43. Reset brake-warning light switch terminal if necessary.

Step 44. Test brakes as required. (Refer to Chapter 15 for instructions.)

Application-Type Automatic Adjuster Brakes. As you perform the following steps refer occasionally to the descriptive drawing of Fig. 9-2.

Step 1. Be sure car is parked on level, hard surface.

Step 2. Place chocks fore and aft of wheel on same side of car as the one to be overhauled. When rear wheel is to be worked on do not set parking brake.

Step 3. Raise corner of car with jack.

Step 4. Place jack stand under axle or suspension. (Refer to Chapter 7 for proper jack points for your car.) NOTE:

Bumper jacks are untrustworthy objects, so never depend on one. If you do not have a jack stand, place something under the frame strong enough to support the weight of the car, such as a green oak log. But do not work beneath your car while it is precariously perched atop anything so tottery as a bumper jack (or as inherently unreliable as any jack).

Step 6. Pry hubcap off wheel; place hubcap upside down nearby to serve as a receptacle for lug nuts and brake parts.

Step 7a. *Rear Wheel Only:*

(1) Loosen and remove lug nuts.

(2) Remove wheel.

(3) Remove drum from brake—it may require a tap or two to free drum from hub flange. In fact, it may be necessary to adjust brake shoes away from drum; if so, refer to Fig. 9-3 or 9-4 for visual instructions.

Step 7b. *Front Wheel Only:*

(1) Pry grease-retainer cup from hub.

(2) Remove cotter pin.

(3) Remove axle nut—turn it counterclockwise with an adjustable wrench while holding wheel with other hand.

(4) Remove grease retainer.

(5) Wiggle wheel and pull out on it at the same time in order to work outer bearing outward on axle (spindle). Remove bearing.

(6) Remove wheel (and outer bearing cup) from spindle.

(7) Remove inner bearing, cup, and grease retainer from spindle.

(8) Place bearings, cups, and retainers in clean solvent or kerosene. *CAUTION:* Such solvents are both flammable and toxic: keep them away from open flames, sparks, and out of the reach of children.

(9) Wash spindle and wheel hub with solvent.

Step 8. Carefully note position of all parts of brake assembly so that they can be installed into their proper positions.

Step 9. Install a wheel-cylinder clamp (Fig. 9-5) if you have one.

Step 10. Remove shoe-return springs (using special wrench, if available).

Step 11. Remove automatic-adjuster cable, cable guide, and adjuster lever.

Step 12. Remove shoe holddown parts.

Step 13. Spread shoes apart at bottom and remove adjuster screw.

Step 14. On rear wheels only, remove parking-brake strut with its antirattle spring.

Step 15. On rear wheels only, remove horseshoe clip and disconnect parking-brake lever from secondary shoe.

Step 16. Spread shoes at top to clear piston links; remove both shoes, adjusting screw, and the lower shoe-to-shoe spring as an assembly (Fig. 9-6).

Step 17. Separate shoes, adjusting screw, and spring.

Step 18 Wash backing plate with water.

Step 19. Check shoe-support points—they should be smooth and free of rough spots. Clean and sand them if necessary, and apply white lubricant.

Step 20. Check the adjusting-screw threads—be sure nut turns freely on screw. If necessary, clean or replace it. Apply white lubricant to threads.

Step 21. Clean all springs, then check them for loss of tension: drop them from four or five feet onto concrete floor or walkway. A good one lands with a thud; a bad one sounds "springy." Replace any bad ones.

Step 22. Check all brake parts; replace any that are badly worn or damaged.

Step 23. Overhaul wheel cylinder (refer to instructions at end of chapter).

Step 24. On rear wheels only, assemble parking brake-lever on new secondary shoe; secure in place with horseshoe clip.

Step 25. Drive or press a new adjuster-lever pin into small hole near lower end of new secondary shoes (Fig. 9-7); install pin from side of shoe web that faces outward when shoe is mounted on car. Be sure pin is fully seated.

Step 26. Assemble new primary and secondary shoes with adjusting screw and spring.

Step 27. Install assembly on backing plate; spread shoes apart at top and fit shoe ends into cylinder links and on anchor pin. Secure shoes with holddown parts.

Step 28. Assemble antirattle spring on parking-brake strut; position strut between parking-brake lever and primary shoe web.

Step 29. Install anchor plate and adjuster-cable loop (so that override spring can be installed with hook facing outward, toward you) on anchor pin.

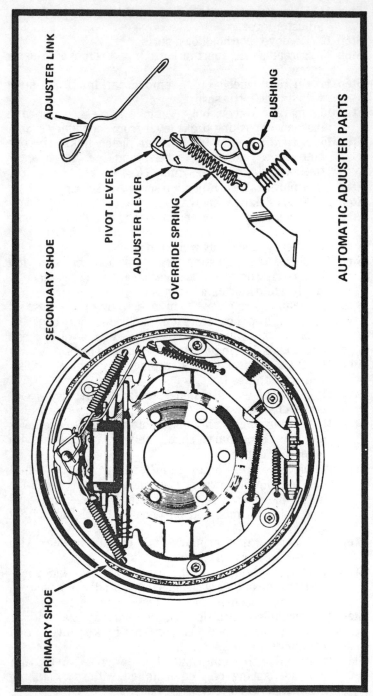

ADJUSTER LINK

BUSHING

PIVOT LEVER

ADJUSTER LEVER

OVERRIDE SPRING

SECONDARY SHOE

PRIMARY SHOE

AUTOMATIC ADJUSTER PARTS

Fig. 9-6. Delco-Moraine duoservo with pivot lever and override spring.

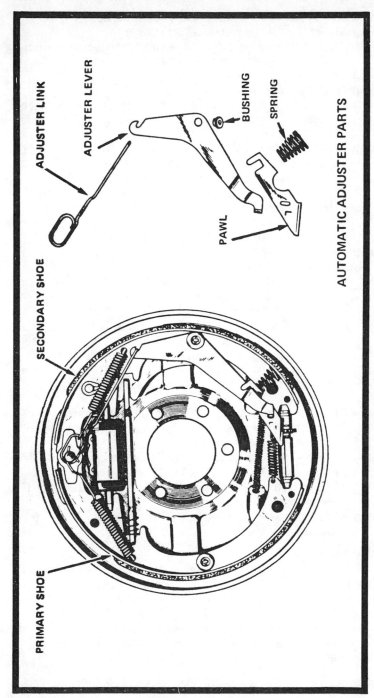

ADJUSTER LINK

ADJUSTER LEVER

BUSHING

SPRING

PAWL

AUTOMATIC ADJUSTER PARTS

SECONDARY SHOE

PRIMARY SHOE

Fig. 9-7. Delco-Moraine duoservo with lever and pawl.

Step 30. If you used a wheel-cylinder clamp, remove it.

Step 31. Install primary-shoe return spring.

Step 32. Position cable guide on secondary-shoe web; then install secondary-shoe return spring.

Step 33. Install adjuster-lever spring.

Step 34. Install adjuster lever.

Step 35. Attach adjuster cable to adjuster lever—be sure override spring is installed with open end of hook facing outward.

Step 36. Inspect brake drum. If necessary, have it turned or replace it. (Refer to Chapter 4 for information on drum damage and repair and replacement data.)

Step 37. Measure or estimate drum diameter; then adjust shoes to about ¼ in. smaller diameter so that drum will slide over easily. Adjust shoes by holding lever clear of star wheel and turning adjusting screw.

Step 38. Install drum over shoes on hub.

   (a) *On Rear Wheel Only*:

   Replace drum flange; then install wheel on lug bolts—tighten lug nuts.

   (b) *On Front Wheel Only*:

   Pack wheel bearings as described at the end of this chapter. Replace retainers, bearings, and wheel on spindle in reverse order of removal (step 7). Tighten wheel nut "hand-tight" with pliers, then back off only as far as necessary to insert cotter pin.

Step 39. Remove jack stand and proceed to next wheel.

Step 40. When all wheels to be overhauled are completed, overhaul master cylinder.

Step 41. Bleed system as required. (Refer to Chapter 12 for instructions.)

Step 42. Reset brake-warning light switch terminal if necessary.

Step 43. Test brakes as required. (Refer to Chapter 15 for instructions.)

Delco-Moraine Duoservo

There are two types of automatic adjuster duoservos employed on Delco-Moraine brakes, pictured in Figs. 9-8 and 9-9. The first type (Fig. 9-8) uses a pivot lever and override spring. For convenience, we'll call it *Type A*. The other (*Type B*) uses a lever and pawl arrangement.

Type A System. As you perform the following steps, refer to Fig. 9-8.

Step 1. Be sure car is parked on level, hard surface.

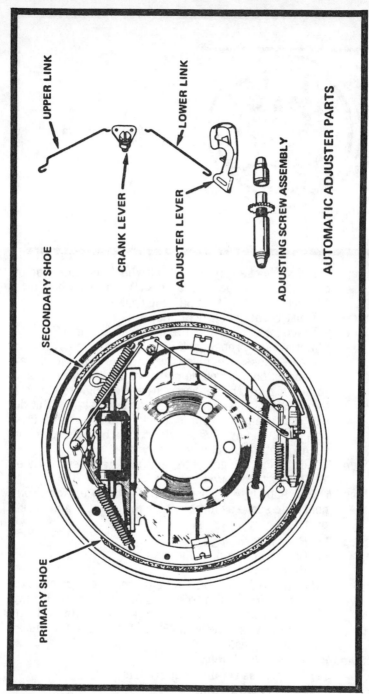

UPPER LINK

CRANK LEVER

LOWER LINK

SECONDARY SHOE

ADJUSTER LEVER

ADJUSTING SCREW ASSEMBLY

PRIMARY SHOE

AUTOMATIC ADJUSTER PARTS

Fig. 9-8. Wagner duoservo with automatic adjust.

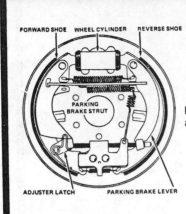

FORWARD SHOE   WHEEL CYLINDER   REVERSE SHOE

PARKING
BRAKE STRUT

ADJUSTER LATCH   PARKING BRAKE LEVER

Fig. 9-9. Bendix nonservo with automatic adjust.

Step 2. Place chocks fore and aft of wheel on same side of car as the one to be overhauled. (When rear wheel is to be worked on do not set parking brake.)

Step 3. Raise corner of car with jack.

Step 4. Place jack stand under axle or suspension. (Refer to Chapter 7 for proper jack points for your car.) *NOTE*: Bumper jacks are untrustworthy objects, so never depend on one. If you do not have a jack stand, place something under the frame strong enough to support the weight of the car, such as a green oak log. But do not work beneath your car while it is precariously perched atop anything so tottery as a bumper jack (or as inherently—unreliable as any jack).

Step 5. Lower car weight onto jack stand; be sure wheel still clears the ground.

Step 6. Pry hubcap off wheel; place hubcap upside down nearby to serve as a receptacle for lug nuts and brake parts.

Step 7a. *Rear Wheel Only:*.
　(1) Loosen and remove lug nuts.
　(2) Remove heel.
　(3) Remove drum from brake—it may require a tap or two to free drum from hub flange. In fact, it may be necessary to adjust brake shoes away from drum; if so, refer to Figs. 9-3 or 9-4 for visual instructions.

Step 7b. *Front Wheel Only:*
　(1) Pry grease-retainer cup from hub.
　(2) Remove cotter pin.

(3) Remove axle nut—turn it counterclockwise with an adjustable wrench while holding wheel with other hand.

(4) Remove grease-retainer cup.

(5) Wiggle wheel and pull out on it at the same time in order to work outer bearing outward on axle (spindle); remove bearing.

(6) Remove wheel (and outer bearing cup) from spindle.

(7) Remove inner bearing, cup, and grease retainer from spindle.

(8) Set bearings, cups, and retainers in solvent or kerosene. NOTES: Such solvents are both flammable and toxic; keep them away from open flames, sparks, and out of reach of children.

(9) Wash spindle and wheel hub with solvent.

Step 8. Carefully note position of all parts of brake assembly so that they can be installed into their proper positions.

Step 9. Install a wheel-cylinder clamp (Fig. 9-5) if you have one.

Step 10. Remove shoe-return springs, (using special tool, if available).

Step 11. Remove automatic-adjuster cable, cable guide, and adjuster lever.

Step 12. Remove shoe holddown parts.

Step 13. Spread shoes apart at bottom and remove adjuster screw.

Step 14. On rear wheels only, remove parking-brake strut with its antirattle spring.

Step 15. On rear wheels only, remove horseshoe clip, and disconnect parking-brake lever from secondary shoe.

Step 16. Spread shoes at top to clear piston links; remove both shoes, adjusting screw, and the lower shoe-to-shoe spring as an assembly (Fig. 9-6).

Step 17. Separate shoes, adjusting screw, and spring.

Step 18. Wash backing plate with water and detergent; dry thoroughly.

Step 19. Check shoe-support points—they should be smooth and free of rough spots. Clean and sand them if necessary, and apply white lubricant.

Step 20. Check the adjusting-screw threads—be sure nut turns freely on screw. If necessary, clean or replace it. Apply white lubricant to threads.

Step 21. Clean all springs then check them for loss of tension: drop them from four or five feet onto concrete floor or walkway—a good one lands with a thud; a bad one sounds "springy." Replace any bad ones.

Step 22. Check all brake parts; replace any that are badly worn or damaged.

Step 23. Overhaul wheel cylinder. (Refer to section at end of chapter for instructions.)

Step 24. On rear wheels only, assemble parking-brake lever on new secondary shoe; secure in place with horseshoe clip.

Step 25. Remove old adjuster-lever bushing and install new one.

Step 26. Assemble new primary and secondary shoes with adjusting screw and spring.

Step 27. Install assembly on backing plate; spread shoes apart at top and fit shoe ends into links and on anchor pin. Secure shoes with holddown parts.

Step 28. Assemble antirattle spring on parking-brake strut; position strut between parking-brake lever and primary-shoe web.

Step 29. Install anchor plate and adjuster-cable loop on anchor pin (so that override spring can be installed with hook facing outward, toward you).

Step 30. If you used a wheel cylinder clamp, remove it.

Step 31. Install primary-shoe return spring.

Step 32. Position cable guide on secondary-shoe web; then install secondary-shoe return spring.

Step 33. Install adjuster-lever spring.

Step 34. Install adjuster lever.

Step 35. Attach adjuster cable to adjuster lever—be sure override spring is installed with open end of hook facing outward.

Step 36. Inspect brake drum. If necessary, have it turned or replace it. (Refer to Chapter 4 for information on drum damage and repair and replacement data.)

Step 37. Measure or approximate drum diameter; then adjust shoes to about ¼ in. smaller diameter so that drum will slide over easily. Adjust shoes by holding adjuster lever clear of star wheel and turning adjusting screw.

Step 38. Install drum over shoes on hub.

    (a) *On Rear Wheels Only*:
        Replace drum on flange; then install wheel on lug bolts—tighten lug nuts.

(**b**) *On Front Wheels Only*:

    Pack wheel bearings as described at the end of this chapter. Replace retainers, bearings, and wheel on spindle in reverse order of removal (step 7). Hand-tighten wheel nut, then back off only as far as necessary to insert cotter pin.

Step 39. Remove jack stand and proceed to next wheel.

Step 40. When all wheels to be overhauled are completed, overhaul master cylinder.

Step 41. Bleed system as required. (Refer to Chapter 12 for instructions.)

Step 42. Reset brake-warning light switch terminal if necessary.

Step 43. Test brakes as required. (Refer to Chapter 15 for instructions.)

Type B System. The Delco-Moraine units that have a lever-and-pawl automatic adjuster (Fig. 9-9) require a different procedure:

Step 1. Be sure car is parked on level, hard surface.

Step 2. Place chocks fore and aft of wheel on same side of car as the one to be overhauled. When rear wheel is to be worked on, do not set parking brake.

Step 3. Raise corner of car with jack.

Step 4. Place jack stand under axle or suspension. (Refer back to Chapter 8 for proper jack points for your car.) NOTE: Bumper jacks are untrustworthy objects, so never depend on one. If you do not have a jack stand, place something under the frame strong enough to support the weight of the car, such as a green oak log. But do not work beneath your car while it is precariously perched atop anything so tottery as a bumper jack (or as inherently unreliable as any jack).

Step 5. Lower car weight onto jack stand; be sure wheel still clears the ground.

Step 6. Pry hubcap off wheel; place hubcap upside down nearby to serve as a receptacle for lug nuts and brake parts.

Step 7a. *Rear Wheel Only*:

    (1) Loosen and remove lug nuts.

    (2) Remove wheel.

    (3) Remove drum from brake—it may require a tap or two to free drum from hub flange. In fact, it may be necessary to adjust brake shoes away from drum; if so, refer to Fig. 9-3 or 9-4 for visual instructions.

Step 7b *Front Wheel Only*:

    (1) Pry grease-retainer cup from hub.

    (2) Remove cotter pin.

    (3) Remove axle nut—turn it counterclockwise with an adjustable wrench while holding wheel with other hand.

    (4) Remove grease-retainer cup.

    (5) Wiggle wheel and pull out on it at the same time in order to work outer bearing outward on axle (spindle); remove bearing.

    (6) Remove wheel (and outer bearing cup) from spindle.

    (7) Remove inner bearing, cup, and grease retainer from spindle.

    (8) Set bearings, cups, and retainers in solvent or kerosene. *NOTE*: Such solvents are both flammable and toxic; keep them away from open flames, sparks, and out of reach of children.

    (9) Wash spindle and wheel hub with solvent.

Step 8. Carefully note position of all parts of brake assembly so that they can be installed in their proper positions.

Step 9. Install a wheel-cylinder clamp (Fig. 9-5) if you have one.

Step 10. Remove shoe-return springs (using special wrench, if available).

Step 11. Remove automatic-adjuster cable, cable guide, and adjuster lever.

Step 12. Remove shoe holddown parts.

Step 13. Spread shoes apart at bottom and remove adjuster screw.

Step 14. On rear wheels only, remove parking-brake strut with its antirattle spring.

Step 15. On rear wheels only, remove horseshoe clip and disconnect parking-brake lever from secondary shoe.

Step 16. Spread shoes at top to clear piston links; remove both shoes, adjusting screw, and the lower shoe-to-shoe spring as an assembly Fig. 9-6).

Step 17. Separate shoes, adjusting screw, and spring.

Step 18. Wash backing plate with water.

Step 19. Check shoe-support points—they should be smooth and free of rough spots. Clean and sand them if necessary, and apply white lubricant.

Step 20. Check the adjusting-screw threads—be sure nut turns freely on screw. If necessary clean or replace it. Apply white lubricant to threads.

Step 21. Clean all springs then check them for loss of tension. Drop them from four or five feet onto concrete floor or walkway: a good one lands with a thud; a bad one sounds "springy." Replace any bad ones.

Step 22. Check all brake parts; replace any that are badly worn or damaged.

Step 23. Overhaul wheel cylinder. (Refer to section at end of chapter for instructions.)

Step 24. On rear wheels only, assemble parking-brake lever onto new secondary shoe; secure in place with horseshoe clip.

Step 25. Remove old adjuster-lever bushing and install new one.

Step 26. Assemble new primary and secondary shoes with adjusting screw and spring.

Step 27. Install assembly on backing plate; spread shoes apart at top and fit shoe ends into cylinder links and on anchor pin. Secure shoes with holddown parts.

Step 28. Assemble antirattle spring on parking-brake strut; position strut between parking-brake lever and primary-shoe web.

Step 29. Install anchor plate and adjuster-cable loop on anchor pin (so that override spring can be installed with hook facing outward, toward you).

Step 30. If you used a wheel-cylinder clamp, remove it.

Step 31. Install primary-shoe return spring.

Step 32. Position cable guide on secondary-shoe web; then install secondary-shoe return spring.

Step 33. Install adjuster lever spring.

Step 34. Install adjuster lever.

Step 35. Attach adjuster cable to adjuster lever—be sure override spring is installed with open end of hook facing outward.

Step 36. Inspect brake drum. If necessary, have it turned or replace it. (Refer to Chapter 4 for drum damage and repair and replacement data.)

Step 37. Measure or estimate drum diameter; then adjust shoes to about ¼ in. smaller diameter so that drum will slide over easily. Adjust shoes by holding adjuster lever clear of star wheel and turning adjusting screw.

Step 38. Install drum over shoes on hub.

(**a**) *On Rear Wheels Only*:

Replace drum on flange; then, assemble wheel on lug bolts—tighten lug nuts.

(**b**) *On Front Wheels Only*:

Pack wheel bearings as instructed below. Replace retainers, bearings, and wheel on spindle in reverse order of removal (step 7). Hand-tighten wheel nut then back off only as far as necessary to insert cotter pin.

Step 39. Remove jack stand and proceed to next wheel.

Step 40. When all wheels to be overhauled are completed, overhaul master cylinder.

Step 41. Bleed system as required. (Refer to Chapter 12 for instructions.)

Step 42. Reset brake-warning light switch terminal if necessary.

Step 43. Test brakes as required. (Refer to Chapter 15 for instructions.)

Wagner Duoservo with Automatic Adjuster

This brake is shown in Fig. 9-10.

Step 1. Be sure car is parked on level, hard surface.

Step 2. Place chocks fore and aft of wheel on same side of car as the one to be overhauled. When rear wheel is to be worked on, do not set parking brake.

Step 3. Raise corner of car with jack.

Step 4. Place jack stand under suspension or steering bracket. (Refer to Chapter 7 for proper jack points for your car.) *NOTE*: Bumper jacks are untrustworthy objects, so never depend on one. If you do not have a jack stand, place something under the frame strong enough to support the weight of the car, such as a green oak log. But do not work beneath your car while

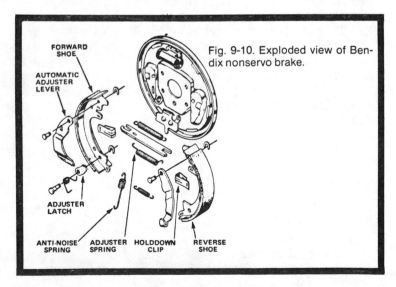

FORWARD SHOE

AUTOMATIC ADJUSTER LEVER

ADJUSTER LATCH

ANTI-NOISE SPRING   ADJUSTER SPRING   HOLDDOWN CLIP   REVERSE SHOE

Fig. 9-10. Exploded view of Bendix nonservo brake.

it is precariously perched atop anything so tottery as a bumper jack (or as inherently unreliable as any jack).

Step 5. Lower car weight onto jack stand; be sure wheel still clears the ground.

Step 6 Pry hubcap off wheel; place hubcap upside down nearby to serve as a receptacle for lug nuts and brake parts.

Step 7a. *Rear Wheel Only:*

    (1) Loosen and remove lug nuts.

    (2) Remove wheel.

    (3) Remove drum from brake—it may require a tap or two to free drum from hub flange. In fact, it may be necessary to adjust brake shoes away from drum; if so, refer to Figs. 9-3 or 9-4 for visual instructions.

7b. *Front Wheel Only:*

    (1) Pry grease-retainer cup from hub.

    (2) Remove cotter pin.

    (3) Remove axle nut—turn it counterclockwise with an adjustable wrench while holding wheel with other hand.

    (4) Remove grease-retainer cup.

    (5) Wiggle wheel and pull out on it at the same time to work outer bearing outward on axle (spindle); remove bearing.

    (6) Remove wheel (and outer bearing cup) from spindle.

    (7) Remove inner bearing, cup, and grease retainer from spindle.

    (8) Set bearings, cups, and retainers in solvent or kerosene. CAUTION: Such solvents are both flammable and toxic: keep them away from open flames, sparks, and out of reach of children.

    (9) Wash spindle and wheel hub with solvent.

Step 8. Carefully note position of all parts of brake assembly so that they can be installed in their proper positions.

Step 9. Install a wheel cylinder clamp (Fig. 9-5) if you have one.

Step 10. Remove shoe-return springs (using special wrench, if available).

Step 11. Remove automatic-adjuster cable, cable guide, and adjuster lever.

Step 12. Remove shoe holddown parts.

Step 13. Spread shoes apart at bottom and remove adjuster screw.

Step 14. On rear wheels only, remove parking-brake strut with its antirattle spring.

Step 15. On rear wheels only, remove horseshoe clip, and disconnect parking-brake lever from secondary shoe.

Step 16. Spread shoes at top to clear piston links; remove both shoes, adjusting screw, and the lower shoe-to-shoe spring as an assembly (Fig. 9-6).

Step 17. Separate shoes, adjusting screw, and spring.

Step 18. Wash backing plate with water.

Step 19. Check shoe-support points—they should be smooth and free of rough spots. Clean and sand them if necessary, and apply white lubricant.

Step 20. Check the adjusting-screw threads—be sure nut turns freely on screw. If necessary, clean or replace it. Apply white lubricant to threads.

Step 21. Clean all springs, then check them for loss of tension: drop them from four or five feet onto concrete floor or walkway—a good one lands with a thud; a bad one sounds "springy." Replace any bad ones.

Step 22. Check all brake parts; replace any that are badly worn or damaged.

Step 23. Overhaul wheel cylinder. (Refer to section at end of chapter for instructions.)

Step 24. *On rear wheels only* assemble parking-brake lever on new secondary shoe; secure in place with horseshoe clip.

Step 25. Remove old adjuster-lever bushing and install new one.

Step 26. Assemble new primary and secondary shoes with adjusting screw and spring.

Step 27. Install assembly on backing plate; spread shoes apart at top and fit shoe ends into cylinder links and anchor pin. Secure shoes with holddown parts.

Step 28. Assemble antirattle spring on parking-brake strut; position strut between parking-brake lever and primary-shoe web.

Step 29. Install anchor plate and adjuster-cable loop on anchor pin (so that override spring can be installed with hook facing outward, toward you).

Step 30. If you used a wheel-cylinder clamp, remove it.

Step 31. Install primary-shoe return spring.

Step 32. Position cable guide on secondary-shoe web; then install secondary-shoe return spring.

Step 33. Install adjuster-lever spring.

Step 34. Install adjuster lever.

Step 35. Attach adjuster cable to adjuster lever—be sure override spring is installed with open end of hook facing outward.

Step 36. Inspect brake drum. If necessary, have it turned or replace it. (Refer to Chapter 4 for information on drum damage and repair replacement data.)

Step 37. Measure or approximate drum diameter; then adjust shoes to about ¼ in. smaller diameter so that drum will slide over easily. Adjust shoes by holding adjuster lever clear of star wheel and turning adjusting screw.

Step 38. Install drum over shoes on hub.

    (**a**) *Rear Wheels Only*:

    Replace drum on flange; then assemble wheel on lug bolts—tighten lug nuts.

    (**b**) *Front Wheels Only*:

    Pack wheel bearings as instructed below. Replace retainers, bearings, and wheel on spindle in reverse order of removal (step 7). Hand-tighten wheel nut, then back off only as far as necessary to insert cotter pin.

Step 39. Remove jack stand and proceed to next wheel.

Step 40. When all wheels to be overhauled are completed, overhaul master cylinder.

Step 41. Bleed system as required. (Refer to Chapter 12 for instructions.)

Step 42. Reset brake-warning light switch terminal if necessary.

Step 43. Test brakes as required. (Refer to Chapter 15 for instructions.)

Nonservo Brakes with Automatic Adjuster

Bendix and Delco-Moraine both produce nonservo brakes with automatic adjusters.

Bendix Type. The step-by-step instructions included here are for rear-wheel brakes and include several references to parking-brake parts. Ignore these references when working with front-wheel brakes. As you perform the sequences, refer occasionally to Fig. 9-11.

Step 1. Be sure car is parked on level, hard surface.

Step 2. Place chocks for and aft of wheel on same side of car as the one to be overhauled. When rear wheel is to be worked on do not set parking brake.

Step 3. Raise corner of car with jack.

Step 4. Place jack stand under axle or suspension. (Refer to Chapter 7 for proper jack points for your car.) NOTE:

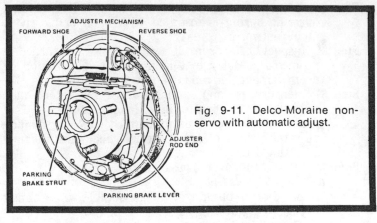

Fig. 9-11. Delco-Moraine non-servo with automatic adjust.

Bumper jacks are untrustworthy objects, so never depend on one. If you do not have a jack stand, place something under the frame strong enough to support the weight of the car, such as a green oak log. But do not work beneath your car while it is precariously perched atop anything so tottery as a bumper jack (or as inherently unreliable as any jack).

Step 5. Lower car weight onto jack stand; be sure wheel still clears the ground.

Step 6. Pry hubcap off wheel; place hubcap upside down nearby to serve as a receptacle for lug nuts and brake parts.

Step 7a. *Rear Wheel Only*:

(1) Loosen and remove lug nuts.
(2) Remove wheel.
(3) Remove drum from brake—it may require a tap or two to free drum from hub flange.

7b. *Front Wheel Only*:

(1) Pry grease retainer from hub.
(2) Remove cotter pin.
(3) Remove axle nut—turn it counterclockwise with an adjustable wrench while holding wheel with other hand.
(4) Remove grease retainer.
(5) Wiggle wheel and pull out on it at the same time to work outer bearing outward on axle (spindle) far enough to grasp it and remove it.
(6) Remove wheel (and outer bearing cup) from spindle.
(7) Remove inner bearing, cup, and grease retainer from spindle.

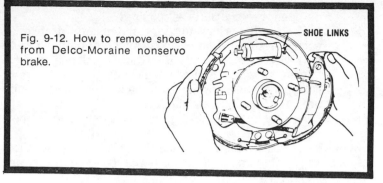

Fig. 9-12. How to remove shoes from Delco-Moraine nonservo brake.

SHOE LINKS

(8) Place bearings, cups, and retainers in clean solvent or kerosene. *CAUTION*: Such solvents are both flammable and toxic: keep them away from open flames, sparks, and out of reach of children.

(9) Wash spindle and wheel hub with solvent.

Step 8. Carefully note position of all parts of brake assembly so that they can be installed in their proper positions.

Step 9. Install wheel cylinder clamp (Fig. 9-5) if you have one.

Step 10. Unhook adjuster spring from parking-brake strut and secondary (reverse) shoe (Fig. 9-12).

Step 11. Unhook upper shoe-to-shoe spring from shoe.

Step 12. Unhook antinoise spring at both ends.

Step 13. Remove parking-brake strut.

Step 14. Disengage both shoes from holddown clips.

Step 15. Unhook lower shoe-to-shoe spring from both shoes.

Step 16. Remove primary (forward) shoe.

Step 17. Disconnect parking-brake cable from parking-brake lever.

Step 18. Remove reverse shoe.

Step 19. Remove shoe holddown clips from backing plate.

Step 20. Remove horseshoe clip from pin and remove parking-brake lever from reverse shoe.

Step 21. Remove horseshoe clip from pin and remove automatic-adjuster lever, latch, and spring from forward shoe.

Step 22. Wash backing plate with water.

Step 23. Overhaul wheel cylinder. (Refer to the section at the end of this chapter for instructions.)

Step 24. Inspect backing plate; be sure brake-support surfaces are clean and smooth—apply white lubricant to these points.

159

Step 25. Inspect all brake parts for excessive wear or damage; replace parts as required.

Step 26. Check all springs for loss of tension: drop each spring from a height of about four feet onto concrete slab. Good springs land with a thud; weak ones make a "springy" sound. Replace weak ones.

Step 27. Apply white brake lubricant to anchor pin, adjuster-lever pin, and latch pin.

Step 28. Assemble latch, spring, and adjuster lever on new forward shoe.

Step 29. Assemble parking-brake lever on reverse shoe. *NOTE*: Reverse shoe lining is narrower than forward shoe lining.

Step 30. Apply white lubricant to inner surface of shoe holddown clips; assemble short side of clips into backing-plate slots.

Step 31. Attach parking-brake cable to parking-brake lever.

Step 32. Hook lower shoe-to-shoe spring into both shoes.

Step 33. Position shoes on backing plate, and slip shoe webs into holddown clips without disengaging clips from backing plate.

Step 34. Install parking-brake strut—tab on strut must engage slot of adjuster lever.

Step 35. Position upper ends of shoe webs into wheel cylinder; install upper shoe-to-shoe spring.

Step 36. If you used wheel-cylinder clamp, remove it.

Step 37. Connect adjuster spring to reverse shoe and into slot in parking-brake strut.

Step 38. Install antinoise spring between reverse shoe and permanent bracket.

Step 39. Inspect drum for damage or excessive wear. Have drum turned or replace it as necessary. (Refer to Chapter 4 for repair and replacement data.)

Step 40. Measure or approximate drum diameter; then adjust shoes to about ¼ in. smaller diameter so that drum will slide over easily. Adjust shoes by holding adjuster lever clear of star wheel and turning screw.

Step 41. Install drum over shoes on hub:

(**a**) *Rear Wheels Only*:
Replace drum on flange; then assemble wheel on lug bolts—tighten lug nuts.

(**b**) *Front Wheels Only*:
Pack wheel bearings as described at the end of this chapter. Replace retainers, bearings, and wheel on spindle in reverse order of removal (step 7). As you turn wheel slowly, tighten wheel nut, then back off only as far as necessary to insert cotter pin.

Step 42. Remove jack stand and proceed to next wheel.

Step 43. When all wheels to be overhauled are completed, overhaul master cylinder.

Step 44. Bleed system as required. (Refer to Chapter 12 for instructions.)

Step 45. Reset brake-warning light switch terminal if necessary.

Step 46. Test brakes as required. (Refer to Chapter 15 for instructions.)

Delco-Moraine. This brake is shown in Fig. 9-13. Check this illustration occasionally during the servicing procedure.

Step 1. Be sure car is parked on level, hard surface.

Step 2. Place chocks fore and aft of wheel on same side of car as the one to be overhauled. When rear wheel is to be worked on, do not set parking brake.

Step 3. Raise corner of car with jack.

Step 4. Place jack stand under axle or suspension. (Refer to Chapter 7 for proper jack points for your car.) *NOTE:* Bumper jacks are untrustworthy objects, so never depend on one. If you do not have a jack stand, place something under the frame strong enough to support the weight of the car, such as a green oak log. But do not work beneath your car while it is precariously perched atop anything so tottery as a bumper jack (or as inherently unreliable as any jack).

Step 5. Lower car weight onto jack stand; be sure wheel still clears the ground.

Step 6. Pry hubcap off wheel; place hubcap upside down nearby to serve as a receptacle for lug nuts and brake parts.

Step 7. (1) Loosen and remove lug nuts.
(2) Remove wheel.

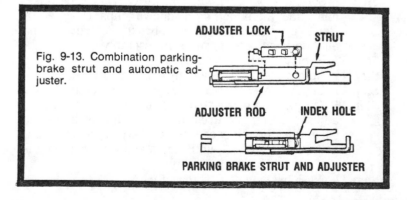

Fig. 9-13. Combination parking-brake strut and automatic adjuster.

ADJUSTER LOCK STRUT

ADJUSTER ROD INDEX HOLE

PARKING BRAKE STRUT AND ADJUSTER

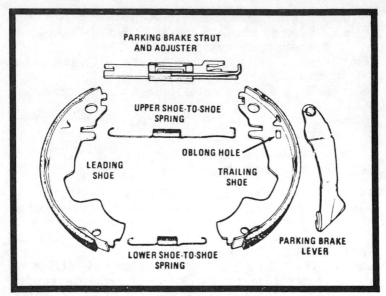

PARKING BRAKE STRUT
AND ADJUSTER

UPPER SHOE-TO-SHOE
SPRING

OBLONG HOLE

LEADING
SHOE

TRAILING
SHOE

PARKING BRAKE
LEVER

LOWER SHOE-TO-SHOE
SPRING

Fig. 9-14. Delco-Moraine nonservo brake showing relationship of parking-brake strut and shoes.

> (3) Remove drum from brake—a tap or two may be required to free drum from hub flange.

Step 8. Carefully note position of all parts of brake assembly so that they can be installed in their proper positions.

Step 9. Adjust nut at parking-brake equalizer to remove all tension from parking-brake cable.

Step 10. Disconnect parking-brake cable from parking-brake lever.

Step 11. Remove upper shoe-to-shoe spring.

Step 12. Spread shoes apart at top, away from shoe links, and slide them out of holddown clips (Fig. 9-14). Remove shoes, parking-brake strut, and automatic-adjuster parts as an assembly.

Step 13. Remove parking-brake strut.

Step 14. Remove adjuster assembly from reverse shoe.

Step 15. Unhook lower shoe-to-shoe spring from both shoes.

Step 16. The parking-brake strut is also the automatic brake adjuster assembly. Press down on adjuster locks and slide adjuster rod over locks to the position shown at bottom of Fig. 9-15—until adjuster lock-index hole is half covered by rod assembly.

Step 17. Wash backing plate with water.

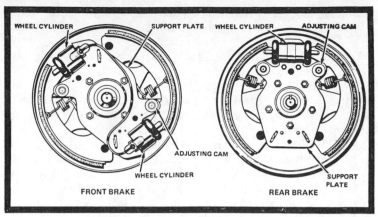

Fig. 9-15. Chrysler center-plane brake.

Step 18. Overhaul wheel cylinder. (Refer to section at end of chapter for instructions.)

Step 19. Check all brake parts for excessive wear or damage, and replace parts if necessary.

Step 20. Check all springs for loss of tension: Drop springs from four or five feet onto concrete slab—good springs should land with a thud; weak springs will make a "springy" sound. Replace weak springs.

Step 21. Apply a small amount of white lubricant to all shoe-support points on backing plate.

Step 22. Assemble parking-brake lever to reverse shoe (Fig. 9-16).

Step 23. Attach parking-brake strut to reverse shoe—strut end fits into oblong hole in shoe web.

Step 24. Hook lower shoe-to-shoe spring into both shoes.

Step 25. Position shoes on backing plate, as shown in Fig. 9-14.

Step 26. Slip shoes into holddown clips. Be sure web ends properly engage shoe links of wheel cylinder pistons.

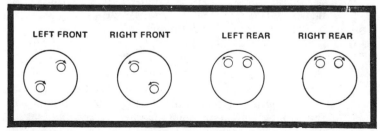

Fig. 9-16. How to adjust Chrysler center-plane brakes—arrows indicate direction to rotate cams to expand shoes as viewed from behind backing plate.

Step 27. Hook upper shoe-to-shoe spring into webs, with square-hook end hooked into reverse-shoe web (Fig. 9-16).

Step 28. Connect parking-brake cable into parking-brake lever. Do not move adjuster lock from position set in step 16.

Step 29. Inspect drum for damage or excessive wear. Have drum turned, or replace it as necessary. (Refer to Chapter 4 for repair and replacement data.)

Step 30. Measure or estimate drum diameter; then adjust shoes to about ¼ in. smaller diameter so that drum will slide over easily. Adjust shoes by holding adjuster lever clear of star wheel and turning adjusting screw.

Step 31. Replace drum on flange; install wheel on lug bolts—tighten lug nuts.

Step 32. Remove jack stand and proceed to next wheel.

Step 33. When all wheels to be overhauled are completed, overhaul master cylinder.

Step 34. Bleed system as required. (Refer to Chapter 12 for instructions.)

Step 35. Reset brake-warning light switch terminal if necessary.

Step 36. Test brakes as required. (Refer to Chapter 15 for instructions.)

> *NOTE*: Set and release parking brake three or four times. If it does not adjust, parking-brake equalizer nuts on rear-wheel brakes should be tightened slightly until adjustment is obtained.

Chrysler Center-Plane Brake

This brake is shown in Fig. 9-15. Refer to this illustration periodically while performing the servicing procedure.

Step 1. Be sure car is parked on level, hard surface.

Step 2. Place chocks fore and aft of wheel on same side of car as the one to be overhauled. When rear wheel is to be worked on, do not set parking brake.

Step 3. Raise corner of car with jack.

Step 4. Place jack stand under axle or suspension. (Refer to Chapter 7 for proper jack points for your car.) *NOTE*: Bumper jacks are untrustworthy objects, so never depend on one. If you do not have a jack stand, place something under the frame strong enough to support the weight of the car, such as a green oak log. But do not work beneath your car while it is precariously perched atop anything so tottery as a bumper jack (or as inherently unreliable as any jack).

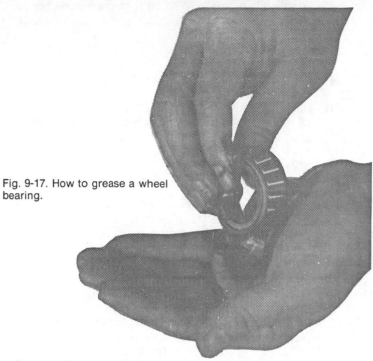

Fig. 9-17. How to grease a wheel bearing.

Step 5. Lower weight onto jack stand; be sure wheel still clears the ground.

Step 6. Pry hubcap off wheel; place hubcap upside down nearby to serve as a receptacle for lug nuts and brake parts.

Step 7a. *Rear Wheel Only*:

(1) Loosen and remove lug nuts.

(2) Remove wheel.

(3) If shoes are pressed against drum, back off the adjusting cams. NOTE: These adjusting cams extend through backing plate for easy access (Fig. 9-18). When viewed from back side of plates, cams should be rotated as arrows indicate to expand shoes, and in the opposite direction to retract shoes. Remove drum.

Step 7b. *Front Wheel Only*:

(2) Pry grease retainer from hub.

(3) Remove axle nut—turn it counterclockwise with adjustable wrench while holding wheel with other hand.

(4) Remove grease retainer.

(5) Wiggle wheel and pull out on it at the same time in order to work the outer bearing outwards on axle (spindle); remove bearing.

(6) Use wheel puller to remove wheels that are keyed to axle.

(7) Remove inner bearing, cup, and grease retainer from spindle.

(8) Place bearings, cups, and retainers in a container of clean solvent or kerosene. *CAUTION*: Such solvents are both flammable and toxic: keep them away from open flames, sparks, and out of reach of children.

(9) Wash spindle and wheel hub with solvent.

Step 8. Carefully note position of all parts of brake assembly so that they can be installed in their proper positions.

Step 9. Install wheel cylinder clamp if you have one (Fig. 9-5).

Step 10. Detach and remove shoe-return springs.

Step 11. Detach and remove both shoes.

Step 12. Wash backing plate and shoe-support plate with water.

Step 13. Overhaul wheel cylinders—refer to section at end of chapter for instructions.

Step 14. Check springs for loss of tension: Drop springs on concrete slab from three or four feet; good springs should land with a thud, while weak springs will make a "springy" sound. Replace weak springs.

Step 15. Apply white brake lubricant to brake-shoe support surfaces.

Step 16. Turn adjusting cams to provide greatest lining-to-drum clearance.

Step 17. Position shoes onto support plate. Be sure shoe webs properly engage wheel cylinder links.

Step 18. Install shoe-return springs.

Step 19. Inspect brake drum. If necessary, have it turned or replace it. (Refer to Chapter 4 for information on drum damage and repair and replacement data.)

Step 20. Measure or estimate drum diameter; then adjust shoes to about ¼ in. smaller diameter so that drum will slide over easily. Adjust shoes by turning adjuster cams as shown in Fig. 9-18.

Step 21. Install drum over shoes on hub.

(a). *Rear Wheels Only*:
Replace drum on flange; then assemble wheel on lug bolts—tighten lug nuts and replace key.

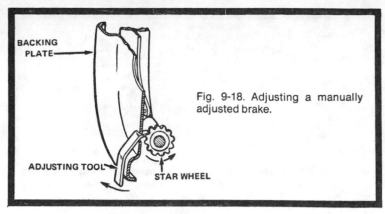

BACKING PLATE

ADJUSTING TOOL

STAR WHEEL

Fig. 9-18. Adjusting a manually adjusted brake.

    (b). *Front Wheels Only*:
    Pack wheel bearings as instructed below. Replace retainers, bearings, and wheel onto spindle in reverse order of removal (step 7). Rotate wheel slowly while tightening wheel nut; back nut off only as far as necessary to insert key (or pin).

Step 22. Adjust brakes: Rotate wheel slowly while tightening adjusting cams one at a time until rotation becomes difficult; then back off the cam just enough to allow wheel to rotate freely once more.

Step 23. Remove jack stand; lower the jack; remove wheel chocks; proceed to next wheel. Repeat steps 2 through 23.

Step 24. When all wheels to be overhauled have been completed, overhaul master cylinder.

Step 25. Bleed system as required. (Refer to Chapter 12 for instructions.)

Step 26. Reset brake-warning light switch terminal, if necessary (not likely to be standard equipment.)

Step 27. Test brakes as required. (Refer to Chapter 15 for instructions.)

HOW TO PACK A WHEEL BEARING

Step 1. Thoroughly clean inner bearing with solvent and allow it to dry.

Step 2. Place wheel-bearing grease in palm of one hand, as shown in Fig. 9-3.

Step 3. Hold bearing in other hand, as shown, and press it into grease. Turn bearing slowly and continue to press it into grease until grease is forced into all parts of bearing and out the top side.

167

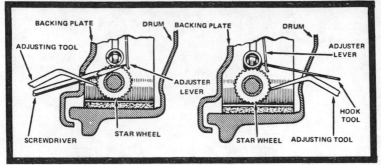

Fig. 9-19. Manually adjusting a brake that has automatic adjuster:
(A) adjusting slots on backing plate;
(B) adjusting slots on drum.

Step 4. Spread grease around outer surface and on all surfaces of cup.

Step 5. Repeat steps 1—4 with outer bearing.

Step 6. Grease the wheel spindle before mounting bearings on it.

HOW TO ADJUST DRUM BRAKES

Since 1957 all American-built cars have used automatic brake adjusters; therefore, the only time you will need to adjust these brakes manually is when an automatic adjuster fails (which is not very often) or after you have worked on a drum-brake assembly.

What You Need

How the automatic adjuster works is explained in Chapter 4. The tools needed for manual adjustment of any drum brake are few and simple. If adjustment is to be made while drum is removed, the adjuster lever can be held clear of the star wheel by hand and a pair of pliers can be used to rotate the adjusting-screw assembly.

If adjustment is to be made while the drum is in place and the wheel mounted, a screwdriver should be used to hold the adjuster lever clear while an adjusting tool is used to rotate the star wheel.

No other tools or materials are needed.

How You Do It

If the drum is removed, hold adjuster lever clear of star wheel. Turn wheel in proper direction to reduce shoe circumference. Reduce circumference enough to allow drum to fit over shoes and on hub.

Manually Adjusted Brakes. After drum and wheel are installed and all lugs have been tightened equally and firmly, with wheel clear of ground, spin wheel slowly with one hand and adjust brake with other hand as shown in Fig. 9-4. Insert adjusting tool through slot in backing plate; expand the shoe (by rotating star wheel) until a heavy drag is felt on the slowly spinning wheel; then back off the star wheel until wheel is just free of that drag.

Automatically Adjusted Brakes. These brakes can be adjusted manually in the same manner as described above, except that the adjusting lever must be held clear of the star wheel to prevent damage to it (Fig. 9-19); however, some of these brakes have the slot in the drum rather than in the backing plate.

Normally, manual adjustment is not necessary. Brakes are quickly adjusted by a few easy forward and backward stops.

HOW TO OVERHAUL A WHEEL CYLINDER

Examine the illustration of Fig. 9-20 first to become generally familiar with the component parts of a typical wheel cylinder. Then perform the following procedure in the sequence shown.

Step 1. Some wheel cylinders must be removed from the backing plate to accommodate disassembly. Disconnect brake-line hose, remove attaching bolts, and—if necessary—remove the bleeder screw.

Step 2. From cylinder bore, remove boot, piston, cup, cup expander (if used), and spring.

Step 3. Thoroughly clean wheel cylinder inside and out with alcohol. **Remember, use no petroleum products**.

Step 4. Inspect cylinder bore for pitting, scratches, burrs, or corrosion.

Step 5. Smooth cylinder bore with crocus cloth. If necessary, use cylinder hone. Use clean brake fluid

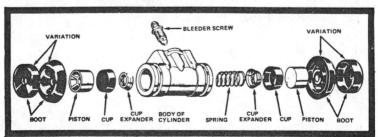

Fig. 9-20. Exploded view of typical wheel cylinder.

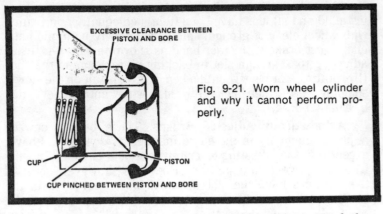

EXCESSIVE CLEARANCE BETWEEN PISTON AND BORE

Fig. 9-21. Worn wheel cylinder and why it cannot perform properly.

CUP

PISTON

CUP PINCHED BETWEEN PISTON AND BORE

to liberally lubricate cylinder bore and crocus cloth or hone while honing. *NOTE*: If cylinder bore does not clean up readily, the wheel cylinder must be replaced.

Step 6. Clean wheel cylinder as in Step 3.

Step 7. Wipe bore dry with a clean, lint-free cloth. Be sure that cylinder, bleeder screw, and brake tube passages are open and free of contaminants of all kinds.

Step 8. Lubricate new piston with clean brake fluid and insert it into cylinder bore; attempt to insert a narrow 0.006 in feeler gage between piston and cylinder wall. If you cannot, proceed to step 9. But if you can, cylinder is too worn to function properly (Fig. 9-21). Throw it away and get a new one.

Step 9. Lubricate interior of cylinder bore with clean brake fluid.

Step 10. Dip rubber parts in clean brake fluid and lay all parts out on a clean sheet of paper.

Step 11. Assemble parts into cylinder bore.

Step 12. If necessary, hold assembly together with wheel-cylinder clamp or a rubber band.

Step 13. If wheel cylinder was removed from backing plate, attach it firmly; install bleeder screw and firmly tighten brake-line hose.

Step-by-Step Disc Brake Overhaul

10

There's no need to be apprehensive about servicing your disc brakes yourself. These brakes are superior to anything you've had previously on a car, and that superiority includes ease of service.

The principles upon which disc brakes operate—and they're much the same as the principles upon which brakes have worked for three-quarters of a century—are discussed in an earlier chapter. The present section is reserved strictly for step-by-step instruction on how to quickly and relatively inexpensively service your disc brakes yourself.

As with drum brakes, two ground rules apply when working on disc brakes: (1) To keep things in balance and insure straight-ahead stops, always overhaul brakes in axle sets. If you do one front wheel, do both front wheels; if you do one rear wheel do both rear wheels. (2) Do one wheel at a time (that is, disassemble, repair, and reassemble one brake) before you touch its axle mate. If you need it, the other will always be there as a pattern, and a pattern can be a useful thing to have at hand, even for the experienced.

WHAT YOU NEED

Relatively few tools and supplies are required:

- Wheel chocks.
- Jack.
- Jack stand (or a reliable substitute).
- Lug wrench (bumper-jack handle usually doubles in this role).
- Box-end or socket wrench of the proper size to remove caliper-mounting bolts.
- Wire clothes hanger to shape into a caliper hanger.
- Replacement plate-and-pad axle set.

171

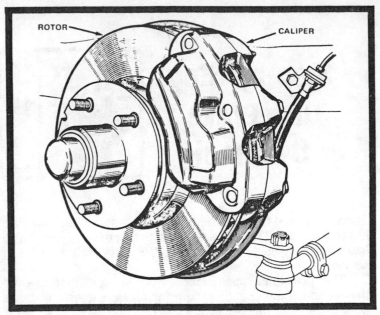

Fig. 10-1. Pre-1968 style disc-brake caliper.

If calipers are to be overhauled, you will need those items listed at the end of this chapter. Furthermore, you should follow any caliper overhaul with a master cylinder overhaul and a careful bleeding of the hydraulic system.

HOW YOU DO IT

Since 1963 several disc-brake designs have been used by the major American car builders. These designs, depicted in Figs. 10-1 (pre-1968), 10-2 (post-1968), and 10-3 are discussed in detail in Chapter 5. Select from those illustrated the type of disc brake your American-built car employs.

After you have selected your brake type, proceed to the appropriate set of step-by-step instructions presented below. Please read through the instructions carefully before purchasing any parts or tools and before attempting any work.

Finally, armed with a good knowledge of what you are about, and the tools and materials you need, start with step 1. You'll be pleasantly surprised at how quickly and easily this work can be completed.

Bendix Four-Piston (Series E) Fixed Caliper

Step 1. Siphon or dip a small amount of brake fluid from master-cylinder reservoir connected to disc brakes;

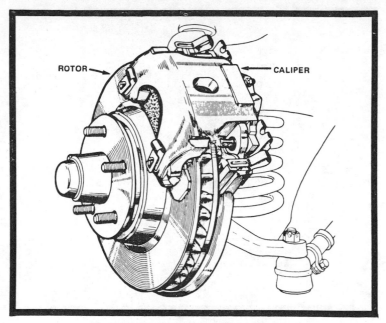

Fig. 10-2. Post-1968 style disc-brake caliper.

otherwise, reservoir may overflow when fluid is forced from caliper cylinders back through line.

Step 2. Be sure your car is parked on a level, hard surface.

Step 3. Place chocks fore and aft of wheel on same side of car as the one being overhauled. When rear wheel is being worked on, do not set parking brake.

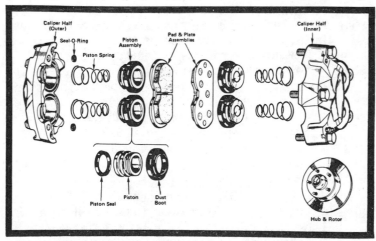

Fig. 10-3. Exploded view of Bendix four-piston caliper.

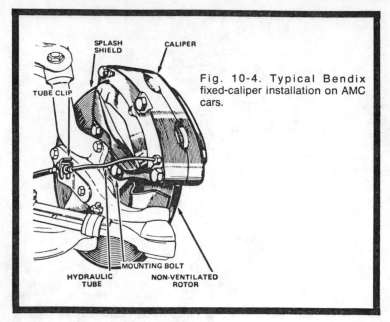

Fig. 10-4. Typical Bendix fixed-caliper installation on AMC cars.

SPLASH SHIELD

CALIPER

TUBE CLIP

HYDRAULIC TUBE

MOUNTING BOLT

NON-VENTILATED ROTOR

Step 4. Raise corner of car with jack.

Step 5. Place jack stand under suspension or steering bracket. (Refer to Chapter 7 for lift point.) NOTE: No jack is trustworthy. If you do not have a jack stand, place something under the axle or frame of your car so that its weight cannot come crashing down on you or on your bared brake rotor.

Step 6. Lower car weight on jack stand, but be sure wheel still clears the ground.

Step 7. Pry hubcap off wheel and place it upside down close by to hold lug nuts and brake parts.

Step 8. Loosen and remove lug nuts and remove wheel.

Step 9. **American Motors**: See Fig. 10-4.

    (1) Remove the two caliper-mounting bolts and shims (note position of shims).

    (2) Remove tube clip from support bracket.

    Dodge and Plymouth: See Fig. 10-5.

    Remove the two caliper-mounting bolts.

    Buick: See Fig. 10-6.

    (1) Remove tube-support bracket.

    (2) Remove caliper-mounting bolts.

Step 10. Lift caliper from rotor and suspend it on a wire hook, as shown in Fig. 10-7 or lay it across tie rod. Do not allow brake hose to support caliper weight.

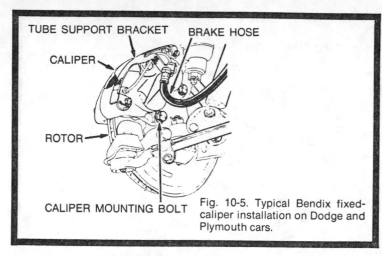

TUBE SUPPORT BRACKET   BRAKE HOSE

CALIPER

ROTOR

CALIPER MOUNTING BOLT   Fig. 10-5. Typical Bendix fixed-caliper installation on Dodge and Plymouth cars.

Step 11. Press pads against pistons—as far as they will go—in order to bottom the pistons in their cylinders.

Step 12. Remove pads and plates from caliper.

Step 13. Inspect caliper for indications of damage or fluid leakage. Overhaul caliper if necessary.

Step 14. Use alcohol and a lint-free cloth to clean all dirt, grime, and rust flakes from caliper. Do not use compressed air to clean caliper—you might unseat boots from pistons. Do not use a petroleum product (such as kerosene) to clean caliper—these products cause rubber parts to swell and are detrimental to pads.

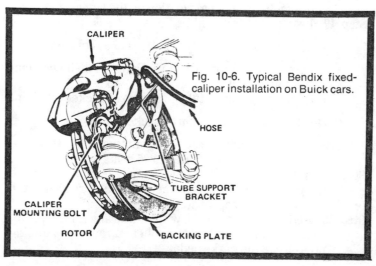

CALIPER

Fig. 10-6. Typical Bendix fixed-caliper installation on Buick cars.

HOSE

TUBE SUPPORT BRACKET

CALIPER MOUNTING BOLT

ROTOR

BACKING PLATE

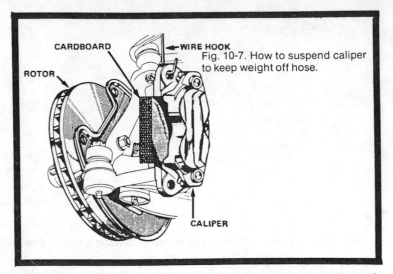

CARDBOARD ROTOR WIRE HOOK CALIPER

Fig. 10-7. How to suspend caliper to keep weight off hose.

Step 15. Inspect all exposed brake parts to be reused. Look for damage or excessive wear. Replace parts as necessary.

Step 16. Inspect rotor. (Refer to Chapter 5 for rotor-damage data and wear tolerances.)

Step 17. Install new pad and plate assemblies into caliper. The metal plates go against the piston ends. Be sure plates are properly seated.

Step 18. Press pads and plates against pistons to be sure they are fully retracted into their cylinders. Then, with pads spread apart in this manner, slide the caliper into its original position on the rotor.

Step 19. Install caliper mounting bolts.

American Motors: Two operations are required.

(1) Be sure shims are installed in their original arrangement.
(2) Attach hydraulic tube to support bracket with clip.

Buick: Install tube support bracket.

Step 20. Insert and tighten caliper-mounting bolts. If torque wrench is available, tighten to 85 ft-lb.

Step 21. Install wheel, remove jack stand, lower and remove the jack, and remove the wheel chocks. Proceed to the next wheel and repeat steps 3 through 21.

Step 22. Refill master cylinder with new brake fluid (specify DOT-3) while assistant applies a gentle pumping pressure to the brake pedal.

Step 23. If calipers were overhauled, bleed brakes. (Refer to Chapter 12 for instructions.)

Step 24. Apply firm brake pressure several times (with engine running for power brakes) to seat pads against rotors.

Step 25. Test brakes. (Refer to chapter 15 for instructions.) *CAUTION:* Do not move car until a firm brake pedal is obtained.

Budd Four-Piston Disc Brake

Step 1. Siphon or dip a small amount of brake fluid from master-cylinder reservoir connected to disc brakes; otherwise, reservoir may overflow when fluid is forced from caliper cylinders back through line.

Step 2. Be sure your car is parked on a level, hard surface.

Step 3. Place chocks fore and aft of wheel on same side of car as the one being overhauled. When rear wheel is being worked on, do not set parking brake.

Step 4. Raise corner of car with jack.

Step 5. Place jack stand under suspension or steering bracket. (Refer to chapter 7 for lift point if you are in doubt as to where to place jack stand.) *NOTE:* No jack is trustworthy. If you do not have a jack stand, place something under the axle or frame of your car so that its weight cannot come crashing down upon you or upon your bared brake rotor.

Step 6. Lower car weight on jack stand, but be sure wheel still clears the ground.

Step 7. Pry hubcap off wheel and place it upside down close by to hold lug nuts and brake parts.

Step 8. Loosen and remove lug nuts and remove wheel.

Step 9. Remove antirattle spring (Figs. 10-8 and 10-9).

Step 10. Remove the two caliper mounting bolts.

Step 11. Lift caliper from rotor and suspend it on a wire hook as shown in Fig. 10-7 or lay it across tie rod. Do not allow brake hose to support caliper weight.

Step 12. Press pads against pistons—as far as they will go—to bottom the pistons in their cylinders.

Step 13. Remove pads and plates from caliper.

Step 14. Inspect caliper for indications of damage or fluid leakage. Overhaul caliper if necessary.

Step 15. Use alcohol and a lint-free cloth to clean all dirt, grime, and rust flakes from caliper. Do not use compressed air to clean caliper—you might unseat boots from pistons. Also, do not use a petroleum

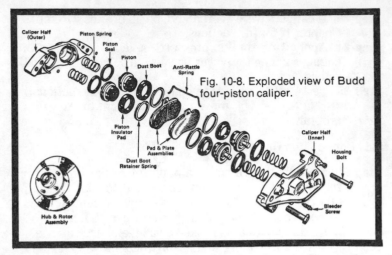

Fig. 10-8. Exploded view of Budd four-piston caliper.

product (such as kerosene) to clean caliper—these products cause rubber parts to swell and are detrimental to pads.

Step 16. Inspect all exposed brake parts to be reused. Look for damage or excessive wear. Replace parts as necessary.

Step 17. Inspect rotor. Refer to Chapter 5 for rotor-damage data and wear tolerances.

Step 18. Install new pad and plate assemblies into caliper. The metal plates go against the piston ends. Be sure plates are properly seated.

Step 19. Press pads and plates against pistons to be sure they are fully retracted into their cylinders. Then,

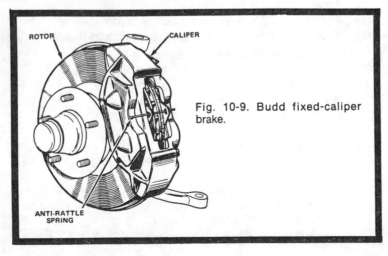

Fig. 10-9. Budd fixed-caliper brake.

with pads spread apart, slide caliper into its original position on the rotor.

Step 20. Install caliper mounting bolts. If torque wrench is available, tighten to 85 ft-lb.

Step 21. Install antirattle spring on caliper.

Step 22. Install wheel; remove jack stand; lower and remove the jack; and remove the wheel chocks: Proceed to next wheel and repeat steps 3 through 22.

Step 23. Refill master cylinder with new brake fluid (specify DOT-3) while assistant applies a gentle pumping pressure to the brake pedal.

Step 24. If calipers were overhauled, bleed brakes. (Refer to chapter 12 for instructions.)

Step 25. Apply firm brake pressure several times (with engine running for power brakes) to seat pads against rotors.

Step 26. Test brakes. (Refer to chapter 15 for instructions.) *CAUTION*: Do not move car until a firm brake pedal is obtained.

Delco-Moraine Four-Piston Caliper

Refer to the illustration of Fig. 10-10 before performing this procedure.

Step 1. Siphon or dip part of brake fluid from master-cylinder reservoir connected to disc brakes; otherwise, reservoir may overflow when fluid is forced from caliper cylinders back through line.

Step 2. Be sure your car is parked on a level, hard surface.

Step 3. Place chocks fore and aft of wheel on same side of car as wheel being overhauled. When rear wheel is being worked on, do not set parking brake.

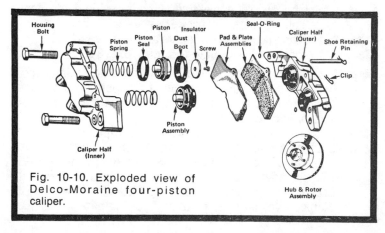

Fig. 10-10. Exploded view of Delco-Moraine four-piston caliper.

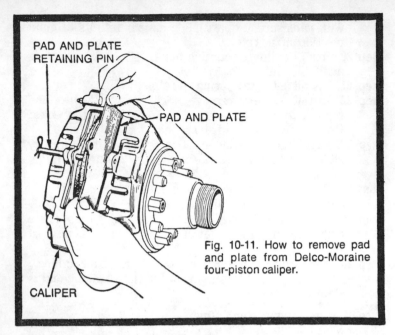

PAD AND PLATE
RETAINING PIN

PAD AND PLATE

Fig. 10-11. How to remove pad and plate from Delco-Moraine four-piston caliper.

CALIPER

Step **4**. Raise corner of car with jack.

Step **5**. Place jack stand under suspension or steering bracket. (Refer to Chapter 7 for lift point if you are in doubt as to where to place jack stand.) *NOTE*: No jack is trustworthy. If you do not have a jack stand, place something under the axle or frame of your car so that its weight cannot come crashing down upon you or upon your bared brake rotor.

Step **6**. Lower car weight on jack stand, but be sure wheel still clears the ground.

Step **7**. Pry hubcap off wheel and place it upside down close by to hold lug nuts and brake parts.

Step **8**. Loosen and remove lug nuts and remove wheel.

Step **9**. Remove pad and plate retaining pin (Fig. 10-11).

Step 10. Insert a screwdriver or brake-adjusting tool between brake rotor and pad; pry apart until pistons are bottomed in caliper cylinders. *CAUTION*: Do not scar or scratch rotor surface.

Step 11. Slip pads and metal plates out of caliper.

Step 12. Inspect caliper for indications of damage or fluid leakage. If necessary, remove and overhaul caliper. (To remove caliper, disconnect brake hose at brake-line support bracket on front caliper or at caliper inlet on rear caliper. Plug or cap brake line and remove caliper-mounting bolts. Lift caliper off rotor.) Refer below for overhaul instructions.

Step 13. Use alcohol and a lint-free cloth to clean all dirt, grime, and rust flakes from caliper. Do not use compressed air to clean caliper—you might unseat boots from pistons. *Do not use a petroleum product* (such as kerosene) *to clean caliper*—these products cause rubber parts to swell and are detrimental to pads.

Step 14. Inspect all exposed brake parts to be reused. Look for damage or excessive wear. Replace parts as necessary.

Step 15. Inspect rotor. (Refer to Chapter 15 for rotor-damage data and wear tolerances.)

Step 16. Install new pad and plate assemblies into caliper. The metal plates fit against the piston ends and the pads fit against the plates. NOTE: If pads and plates are marked with an arrow, be sure they are installed with arrow pointing toward front of car.

Step 17. Install pad-and-plate retaining pin.

Step 18. Install wheel; remove jack stand; lower and remove the jack; and remove the wheel chocks. Proceed to next wheel and repeat steps 3 through 18.

Step 19. Refill master cylinder with new brake fluid (specify DOT-3) while assistant applies a gentle pumping pressure to the brake pedal.

Step 20. If calipers were overhauled, bleed brakes. (Refer to Chapter 12 for instructions.)

Step 21. Apply firm brake pressure several times (with engine running for power brakes) to seat pads against rotors.

Step 22. Test brakes. (Refer to Chapter 15 for instructions.) CAUTION: Do not move car until a firm brake pedal is obtained.

Kelsey-Hayes Four-Piston Caliper (See Fig. 10-12.)

Step 1. Siphon or dip a small amount of brake fluid from master-cylinder reservoir connected to disc brakes; otherwise, reservoir may overflow when fluid is forced from caliper cylinder back through line.

Step 2. Be sure your car is parked on a level, hard surface.

Step 3. Place chocks fore and aft of wheel on same side of car as the one being overhauled. When rear wheel is being worked on, do not set parking brake.

Step 4. Raise corner of car with jack.

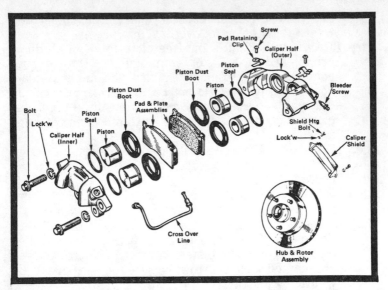

Fig. 10-12. Exploded view of Kelsey-Hayes four-piston caliper.

Step 5. Place jack stand under suspension or steering bracket. (Refer to Chapter 7 for lift point if you are in doubt as to where to place jack stand.) *NOTE*: No jack is trustworthy. If you do not have a jack stand, place something under the axle or frame of your car so that its weight cannot come crashing down upon you or upon your bared brake rotor.

Step 6. Lower car weight on jack stand, but be sure wheel still clears the ground.

Step 7. Pry hub cap off wheel and place it upside down close by to hold lug nuts and brake parts.

Step 8. Loosen and remove lug nuts and remove wheel.

Step 9. Remove pad-retaining clips from caliper (Fig. 10-13).

Step 10. Insert a screwdriver or brake-adjusting tool between brake rotor and pad; pry apart until pistons are bottomed in caliper cylinders. *CAUTION*: Do not scar or scratch rotor surface.

Step 11. Slip pads and metal plates out of caliper.

Step 12. Inspect caliper for indications of damage or fluid leakage. If necessary remove and overhaul caliper. (To remove caliper, disconnect brake hose at frame bracket, plug or cap brake hose, remove caliper-mounting bolts, and lift caliper off rotor.)

Step 13. Use alcohol and a lint-free cloth to clean all dirt, grime, and rust flakes from caliper. Do not use

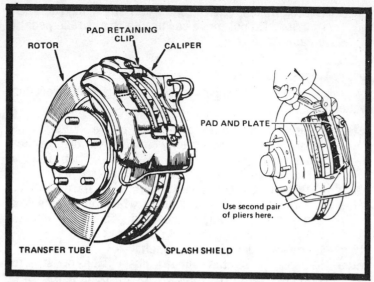

Fig. 10-13. How to remove pad-retaining clips and pad-and-plate assembly from a Kelsey-Hayes four-piston caliper.

compressed air to clean caliper—you might unseat boots from pistons. *Do not use a petroleum product* (such as kerosene) *to clean caliper*—these products cause rubber parts to swell and are detrimental to pads.

Step 14. Inspect all exposed parts to be reused. Look for damage or excessive wear. Replace parts as necessary.

Step 15. Inspect rotor. (Refer to Chapter 5 for rotor-damage data and wear tolerances.)

Step 16. Install new pad-and-plate assemblies into caliper. The metal plates fit against the piston ends. Be sure plate and pad assemblies are properly seated.

Step 17. Install pad-retaining clips on caliper.

Step 18. Install wheel; remove jack stand; lower and remove the jack; and remove the wheel chocks. Proceed to next wheel and repeat steps 3 through 18.

Step 19. Refill master cylinder with new brake fluid (specify DOT-3) while assistant applies gentle pumping pressure to the brake pedal.

Step 20. If calipers were overhauled, bleed brakes. (Refer to Chapter 12 for instructions.)

Step 21. Apply firm brake pressure several times (with engine running for power brakes) to seat pads against rotors.

Step 22. Test brakes. (Refer to Chapter 15 for instructions.) *CAUTION*: Do not move car until a firm brake pedal is obtained.

Kelsey-Hayes (Ford) Floating-Caliper Disc Brake

Refer to Fig. 10-14 before performing this procedure sequence.

Step 1. Siphon or dip part of brake fluid from master-cylinder reservoir connected to disc brakes; otherwise, reservoir may overflow when fluid is forced from caliper cylinder back through line.

Step 2. Be sure your car is parked on a level, hard surface.

Step 3. Place chocks fore and aft of wheel on same side of car as the one being overhauled. When rear wheel is being worked on, do not set parking brake.

Step 4. Raise corner of car with jack.

Step 5. Place jack stand under suspension or steering bracket. (Refer to Chapter 7 for lift point if you are in doubt as to where to place jack stand.) *NOTE*: No jack is trustworthy. If you do not have a jack stand, place something under the axle or frame of your car so that its weight cannot come crashing down upon you or upon your bared brake rotor.

Step 6. Lower car weight on jack stand, but be sure wheel still clears the ground.

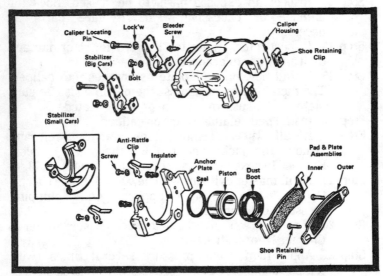

Fig. 10-14. Exploded view of Kelsey-Hayes floating caliper.

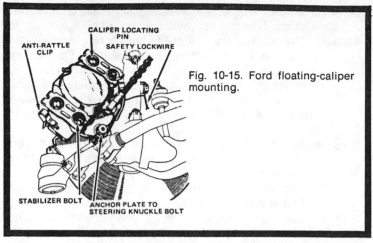

CALIPER LOCATING
PIN
ANTI-RATTLE
CLIP
SAFETY LOCKWIRE

Fig. 10-15. Ford floating-caliper mounting.

STABILIZER BOLT   ANCHOR PLATE TO
STEERING KNUCKLE BOLT

Step   7. Pry hubcap off wheel and place it upside down close by to hold lug nuts and brake parts.

Step   8. Loosen and remove lug nuts and remove wheel.

Step   9. Remove caliper locating pins (Fig. 10-15).

Step 10. Loosen stabilizer bolts and lift caliper off rotor. Suspend caliper from wire hook (Fig. 10-7) or lay it across tie rod. Do not allow brake hose to support weight of caliper.

Step 11. Remove pads and plates from caliper.

Step 12. Remove caliper locating-pin insulators from anchor plate (Fig. 10-16).

Step 13. Gradually press piston into caliper cylinder. Use C-clamp if necessary.

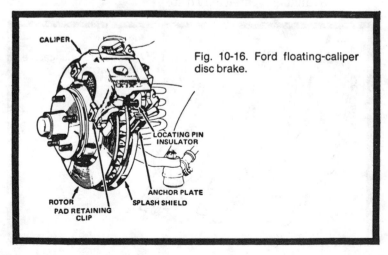

CALIPER

Fig. 10-16. Ford floating-caliper disc brake.

LOCATING PIN
INSULATOR

ANCHOR PLATE

ROTOR
PAD RETAINING
CLIP

SPLASH SHIELD

Step 14. Inspect caliper for indications of damage or fluid leakage. If necessary, overhaul caliper.

Step 15. Use alcohol and a lint-free cloth to clean all dirt, grime, and rust flakes from caliper. Do not use compressed air to clean caliper—you might unseat boot from piston. Do *not use a petroleum product* (such as kerosene) *to clean caliper*—these products cause rubber parts to swell and are detrimental to pads.

Step 16. Inspect all exposed brake parts to be reused. Look for damage or excessive wear. Replace parts as necessary.

Step 17. Inspect rotor. (Refer to Chapter 5 for rotor-damage data and wear tolerances.)

Step 18. Install new caliper locating-pin insulators in anchor plate.

Step 19. Install new pad and plate assemblies on anchor plate; plate tabs must be tucked under antirattle clips. *NOTE*: Three types of pads are illustrated in Fig. 10-17; these are interchangeable.

Step 20. Install new pad-and-plate assembly in outboard side of caliper; fasten in place with holddown pins and retaining clips. If necessary, use a short bolt to hold pins in place while you install clips.

Step 21. Position caliper assembly in place on anchor plate; install caliper locating pins. *NOTE*: To make installation easier, moisten pins with water—do not use oil or grease.

Step 22. Tighten stabilizer bolts. If torque wrench is available, tighten to 10 ft-lb.

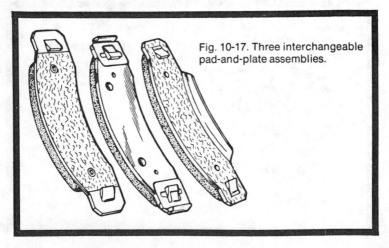

Fig. 10-17. Three interchangeable pad-and-plate assemblies.

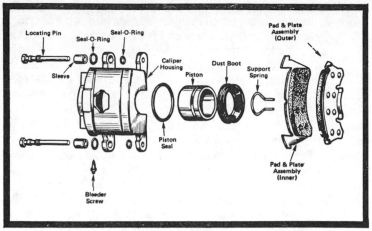

Fig. 10-18. Exploded view of Delco-Moraine floating caliper.

Step 23. Tighten caliper locating pins. If torque wrench is available, tighten to 30 ft-lb.

Step 24. Install wheel; remove jack stand; lower and remove jack; and remove the wheel chocks. Proceed to next wheel and repeat steps 3 through 24.

Step 25. Refill master cylinder with new brake fluid (specify DOT-3) while assistant applies a gentle pumping pressure to the brake pedal.

Step 26. If calipers were overhauled, bleed brakes. (Refer to Chapter 12 for instructions.)

Step 27. Apply firm brake pedal pressure several times (with engine running for power brakes) to seat pads against rotors.

Step 28. Test brakes. (Refer to Chapter 15 for instructions.) CAUTION: Do not move car until a firm brake pedal is obtained.

Delco-Moraine Floating-Caliper Disc Brake

Refer to Fig. 10-18 before performing the steps listed below.

Step 1. Siphon or dip part of brake fluid from master-cylinder reservoir connected to disc brakes; otherwise, reservoir may overflow when fluid is forced from caliper cylinder back through line.

Step 2. Be sure your car is parked on a level, hard surface.

Step 3. Place chocks fore and aft of wheel on same side of car as the one being overhauled. When rear wheel is being worked on, do not set parking brake.

Step 4. Raise corner of car with jack.

Step 5. Place jack stand under suspension or steering bracket. (Refer to Chapter 7 for lift point if you are in doubt as to where to place jack stand.) NOTE: No jack is trustworthy. If you do not have a jack stand, place something under the axle or frame of your car so that its weight cannot come crashing down upon you or upon your bared brake rotor.

Step 6. Lower car weight on jack stand, but be sure wheels still clear the ground.

Step 7. Pry hubcap off wheel and place it upside down close by to hold lug nuts and brake parts.

Step 8. Loosen and remove lug nuts and remove wheel.

Step 9. Place a C-clamp on caliper as shown in Fig. 10-19. Tighten clamp until caliper pistons are bottomed in their cylinders. Remove C-clamp.

Step 10. Remove the two caliper-to-anchor locating pins. Lift caliper off rotor; suspend caliper from wire hook as shown in Fig. 10-7 or lay it across the tie rod. Do not allow brake hose to support weight of caliper.

Step 11. Slip pad and plate assemblies out of caliper.

Step 12. Remove pad-support spring from inside piston. (Refer to Fig. 10-20.)

Step 13. Remove bushings and O-rings from inside caliper ears.

Step 14. Inspect caliper for indications of damage or fluid leakage. If necessary, overhaul caliper. To remove caliper, detach hose from tube, then plug or cap tube.

Step 15. Use alcohol and a lint-free cloth to clean all dirt, grime, and rust flakes from caliper. Do not use compressed air to clean caliper—you might unseat

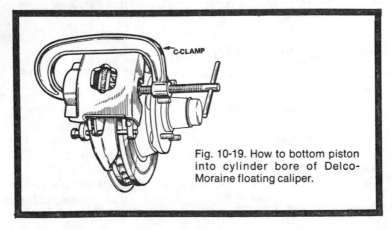

Fig. 10-19. How to bottom piston into cylinder bore of Delco-Moraine floating caliper.

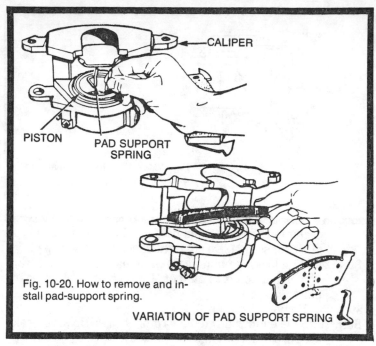

Fig. 10-20. How to remove and install pad-support spring.

VARIATION OF PAD SUPPORT SPRING

boot from piston. Do *not use a petroleum product* (such as kerosene) *to clean caliper*—these products cause rubber parts to swell and are detrimental to pads.

Step 16. Inspect all exposed brake parts to be reused. Look for damage or excessive wear. Replace parts as necessary.

Step 17. Inspect rotor. (Refer to Chapter 7 for rotor-damage data and wear tolerances.)

Step 18. Apply silicone lubricant (or water) to O-rings and to O-ring grooves inside caliper ears. Install O-rings.

Step 19. Apply silicone lubricant (or water) to bushings. Install both bushings. *NOTE*: Ends of bushings should be even with inboard side of ear.

Step 20. Install pad-support spring into piston.

Step 21. Install pad and plate assemblies into caliper. Plate fits against piston end, bottom side of plate against pad-support spring.

Step 22. Rest caliper on rotor, bottom edge of outboard pad on outer edge of rotor in such a way that outboard pad is held firmly against caliper.

Step 23. Place a metal bar (a wrench will do) across the caliper cutout as shown in Fig. 10-21, and install a

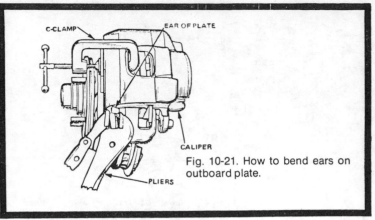

Fig. 10-21. How to bend ears on outboard plate.

C-clamp as shown—this holds pad and plate snugly in place for step 24. Do not tighten enough to scar pad.

Step 24. With locking pliers, grip plate ear about ¼ in. from end and clinch it to caliper. Do both ends of outboard plate.

Step 25. Remove C-clamp.

Step 26. Slide caliper down over rotor. NOTE: Be sure brake hose is neither twisted nor kinked.

Step 27. Install caliper-to-anchor locating pins. NOTE: A bit of silicone lubricant or water on pin ends may make installation easier. Tighten.

Step 28. Install wheel; remove jack stand; lower and remove the jack; and remove the wheel chocks. Proceed to next wheel and repeat steps 3 through 26.

Step 29. Refill master cylinder with new brake fluid (specify DOT-3) while assistant applies a gentle pumping pressure to the brake pedal.

Step 30. If calipers were overhauled, bleed brakes. (Refer to Chapter 12 for instructions.)

Step 31. Apply firm brake-pedal pressure several times (with engine running for power brakes) to seat pads against rotors.

Step 32. Test brakes. (Refer to Chapter 15 for instructions.) CAUTION: Do not move car until a firm brake pedal is obtained.

Kelsey-Hayes (Chrysler and AMC) Floating Caliper Disc.

Refer to Fig. 10-22 before performing the sequence of operations described here.

Step 1. Siphon or dip a small amount of brake fluid from master-cylinder reservoir connected to disc brakes;

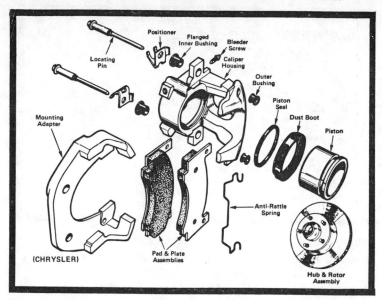

Fig. 10-22. Exploded view of Kelsey-Hayes (Chrysler and AMC) floating caliper.

otherwise, reservoir may overflow when fluid is forced from caliper cylinder back through line.

Step 2. Be sure your car is parked on a level, hard surface.

Step 3. Place chocks fore and aft of wheel on same side of car as the one being overhauled. When rear wheel is being worked on, do not set parking brake.

Step 4. Raise corner of car with jack.

Step 5. Place jack stand under suspension or steering bracket. (Refer to Chapter 7 for lift point if you are in doubt as to where to place jack stand.) NOTE: No jack is trustworthy. If you do not have a jack stand, place something under the axle or frame of your car so that its weight cannot come crashing down upon you or upon your bared brake rotor.

Step 6. Lower car weight on jack stand, but be sure wheel still clears the ground.

Step 7. Pry hubcap off the wheel and place it upside down close by to hold lug nuts and brake parts.

Step 8. Loosen and remove lug nuts and remove wheel.

Step 9. Remove the caliper-to-anchor locating pins. (Fig. 10-23.)

Step 10. Lift caliper off rotor.

Step 11. Support caliper with a wire hanger. Do not allow brake hose to support weight of caliper.

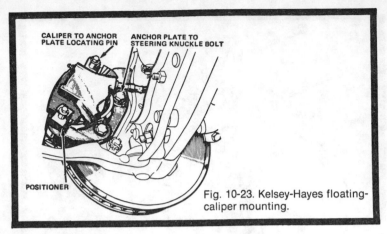

Fig. 10-23. Kelsey-Hayes floating-caliper mounting.

Step 12. Remove antirattle spring (Fig. 10-24.)

Step 13. Slide outboard pad-and-plate assembly out of caliper (Fig. 10-25.)

Step 14. Slide inboard pad-and-plate assembly out of caliper. (Fig. 10-26.)

Step 15. Remove the four bushings from the locating-pin holes in caliper.

Step 16. Inspect caliper for indications of damage or fluid leakage. If necessary overhaul caliper. To remove caliper, disconnect the brake hose from the brake-line tube at the frame-mounting bracket. Plug or cap the tube.

Step 17. Use alcohol and a lint-free cloth to clean all dirt, grime, and rust flakes from caliper. Do not use

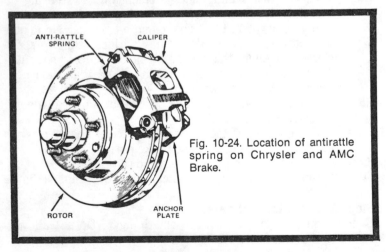

Fig. 10-24. Location of antirattle spring on Chrysler and AMC Brake.

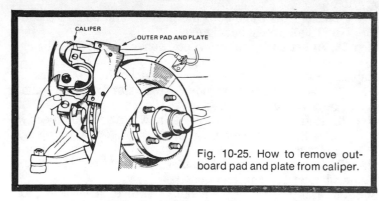

Fig. 10-25. How to remove outboard pad and plate from caliper.

compressed air to clean caliper—you might unseat boot from piston. Do *not use a petroleum product* (such as kerosene) *to clean caliper*—these products cause rubber parts to swell and are detrimental to pads.

Step 18. Inspect all exposed brake parts to be reused. Look for damage or excessive wear. Replace parts as necessary.

Step 19. Inspect rotor. (Refer to Chapter 5 for rotor-damage data and wear tolerances.)

Step 20. Gradually press in on caliper piston until it is bottomed in caliper-cylinder bore. *NOTE*: Use C-clamp if necessary; however, if piston is that hard to bottom, you may have a problem within the hydraulic system itself.

Step 21. Install four new bushings in caliper ears.

Step 22. Install new inboard plate and pad into caliper.

Step 23. Install new outboard plate and pad into caliper; fasten with antirattle spring.

Step 24. Slide caliper over rotor and into position against anchor plate. *NOTE*: Locating-pin holes in caliper,

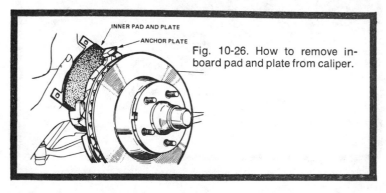

Fig. 10-26. How to remove inboard pad and plate from caliper.

anchor plate, and both the inboard and outboard pad and plates must be aligned.

Step 25. Insert each pin into a positioner. *NOTE*: Positioner arrow must point upward.

Step 26. Install and tighten caliper-to-anchor locating pins.

Step 27. Install wheel; remove jack stand; lower and remove the jack; and remove the wheel chocks.

Step 28. Refill master cylinder with new brake fluid (specify DOT-3) while assistant applies a gentle pumping pressure to brake pedal.

Step 29. If calipers were overhauled, bleed brakes. (Refer to Chapter 15 for instructions.)

Step 30. Apply firm brake-pedal pressure several times (with engine running for power brakes) to seat pads against rotors.

Step 31. Test brakes. (Refer to Chapter 15 for instructions.) *CAUTION*: Do not move car until a firm brake pedal is obtained.

Bendix CA Floating-Caliper Disc

Refer to Fig. 10-27 before performing the service sequence outlined here.

Step 1.Siphon or dip a small amount of brake fluid from master-cylinder reservoir connected to disc brakes;

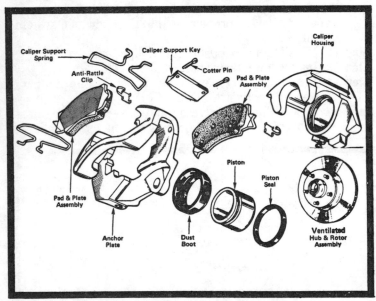

Fig. 10-27. Exploded view of Bendix CA floating caliper.

otherwise, this reservoir may overflow when fluid is forced from caliper cylinder back through line.

Step 2. Be sure your car is parked on a level, hard surface.

Step 3. Place chocks fore and aft of wheel on same side of car as the one being overhauled. When rear wheel is being worked on, do not set parking brake.

Step 4. Raise corner of car with jack.

Step 5. Place jack stand under suspension or steering bracket. (Refer to Chapter 7 for lift point if you are in doubt as to where to place jack stand.) *NOTE*: No jack is trustworthy. If you do not have a jack stand, place something under the axle or frame of your car so that its weight cannot come crashing down upon you or upon your bared brake rotor.

Step 6. Lower car weight on jack stand, but be sure wheel still clears the ground.

Step 7. Pry hubcap off wheel and place it upside down close by to hold lug nuts and brake parts.

Step 8. Loosen and remove lug nuts and remove wheel.

Step 9. Remove cotter pin from caliper-support key (Fig. 10-28) and caliper-support spring (if used).

Step 10. Use a drift or screwdriver and hammer to force caliper-support key from between caliper and anchor plate—this step is illustrated in Fig. 10-29.

Step 11. Lift caliper off anchor plate and clear of rotor. Suspend it on wire hanger (Fig. 10-7) or lay it across tie rod.

Step 12. Remove pads and plates from calipers as shown in Fig. 10-30.

Step 13. Inspect caliper for indications of damage or fluid leakage. If necessary, overhaul caliper. To remove

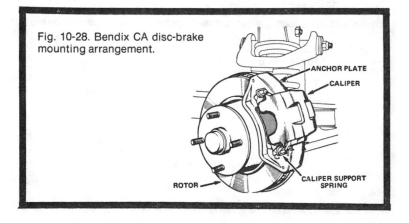

Fig. 10-28. Bendix CA disc-brake mounting arrangement.

ANCHOR PLATE

CALIPER

ROTOR

CALIPER SUPPORT SPRING

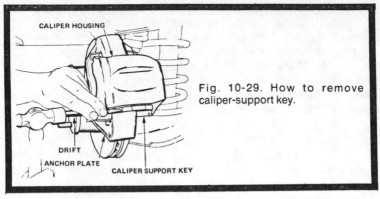

Fig. 10-29. How to remove caliper-support key.

CALIPER HOUSING

DRIFT

ANCHOR PLATE

CALIPER SUPPORT KEY

caliper, disconnect brake hose from caliper inlet. Cap or plug hose.

Step 14. Use alcohol and a lint-free cloth to clean all dirt, grime, and rust flakes from caliper. Do not use compressed air to clean caliper—you might unseat boot from piston. Do *not use a petroleum product* (such as kerosene) *to clean caliper*—these products cause rubber parts to swell and are detrimental to pads.

Step 15. Inspect all exposed brake parts for damage or excessive wear. Replace parts as necessary.

Step 16. Inspect rotor. (Refer to Chapter 5 for rotor-damage data and wear tolerances.)

Step 17. Be sure antirattle clips are properly positioned—loop ends of clips should point away from rotor.

Step 18. Force caliper piston to bottom in its cylinder—hand pressure should be adequate; if it is not, use a C-clamp.

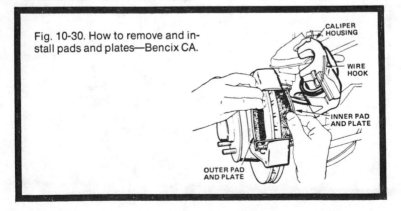

Fig. 10-30. How to remove and install pads and plates—Bencix CA.

CALIPER HOUSING

WIRE HOOK

INNER PAD AND PLATE

OUTER PAD AND PLATE

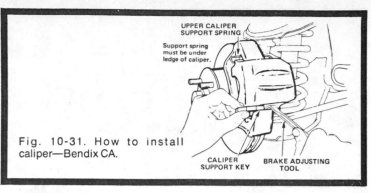

Fig. 10-31. How to install caliper—Bendix CA.

UPPER CALIPER SUPPORT SPRING

Support spring must be under ledge of caliper.

CALIPER SUPPORT KEY

BRAKE ADJUSTING TOOL

Step 19. Insert plate-and-pad assemblies into caliper—remember, the pads must rub against the rotor.

Step 20. Apply a light coating of white lubricant to the anchor plate and caliper surfaces that are in contact with one another after installation. *CAUTION*: Do not allow lubricant to get on pads or the rotor. If it does, remove it as thoroughly as possible.

Step 21. Position caliper on anchor plate (refer to Fig. 10-31)—be sure the caliper-support spring is properly positioned beneath top edge of caliper.

Step 22. Use a screwdriver or brake-adjusting tool (see Fig. 10-31) to hold caliper in place. Insert caliper support key between caliper and anchor plate.

Step 23. Position the caliper support key and install new cotter pins, one in each end.

Step 24. Install wheel; remove jack stand; lower and remove the jack; and remove the wheel chocks. Proceed to next wheel and repeat steps 3 through 24.

Step 25. Refill master cylinder with new brake fluid (specify DOT-3) while assistant applies a gentle pumping pressure to the brake pedal.

Step 26. If calipers were overhauled, bleed brakes. (Refer to Chapter 12 for instructions.)

Step 27. Apply firm brake-pedal pressure several times (with engine running for power brakes) to seat pads against rotors.

Step 28. Test brakes. (Refer to Chapter 15 for instructions.) *CAUTION*: Do not move car until a firm brake pedal is obtained.

Vega Floating-Caliper Disc Brake

Refer to Fig. 10-32 before performing the following service procedure.

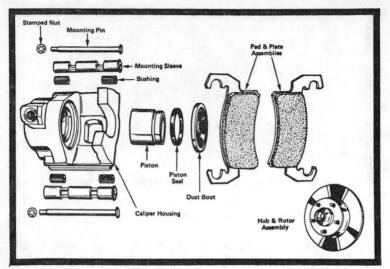

Fig. 10-32. Exploded view of Vega floating caliper.

Step 1. Siphon or dip a small amount of brake fluid from master-cylinder reservoir connected to disc brakes.

Step 2. Be sure your car is parked on a level, hard surface.

Step 3. Place chocks fore and aft of wheel on same side of car as the one being overhauled.

Step 4. Raise corner of car with jack.

Step 5. Place jack stand under suspension or steering bracket. (Refer to Chapter 7 for lift point if you are in doubt as to where to place jack stand.) *NOTE*: If you do not have a jack stand, place something under the axle or frame of your car so that its weight cannot come crashing down upon you or your bared brake rotor.

Step 6. Lower car weight on jack stand, but be sure wheel still clears the ground.

Step 7. Pry hubcap off wheel and place it upside down close by to hold lug nuts and brake parts.

Step 8. Loosen and remove lug nuts and remove wheel.

Step 9. Remove stamped nut from inboard end of each caliper mounting pin (Fig. 10-33).

Step 10. Remove the two caliper mounting pins (Fig. 10-34).

Step 11. Lift caliper off anchor plate but not off rotor.

Step 12. Slide the outboard pad and plate through opening in mounting sleeve and out of caliper.

Step 13. Slide the inboard pad and plate through opening in mounting sleeve and out of caliper.

Step 14. Remove both mounting sleeves.

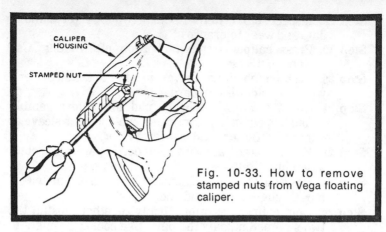

Fig. 10-33. How to remove stamped nuts from Vega floating caliper.

Step 15. Inspect caliper for indications of damage or fluid leakage. If necessary, overhaul caliper. To remove caliper, disconnect the brake hose from the tube. Cap or plug the tube.

Step 16. Use alcohol and a lint-free cloth to clean all dirt, grime, and rust flakes from caliper. Do not use compressed air to clean caliper—you might unseat piston. Do *not use a petroleum product* (such as kerosene) *to clean caliper*—these products cause rubber parts to swell and are detrimental to brake pads.

Step 17. Inspect all exposed brake parts to be reused. Look for damage or excessive wear. Replace parts as necessary.

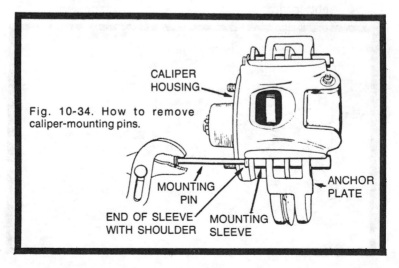

Fig. 10-34. How to remove caliper-mounting pins.

Step 18. Inspect rotor. (Refer to Chapter 5 for rotor-damage data and wear tolerances.)

Step 19. Press caliper piston to bottom of cylinder. Use C-clamp if necessary.

Step 20. Position mounting sleeves in caliper; the sleeve end with a shoulder is outboard.

Step 21. Insert end-tab of inboard pad-and-plate assembly through opening in mounting sleeve and over sleeve at either end. Be sure pad is toward rotor.

Step 22. Repeat step 21 with outboard pad and plate assembly. (Both assemblies are the same.)

Step 23. Position caliper on anchor plate; insert mounting pins through sleeves and anchor plate.

Step 24. Hold caliper assembly firmly together; install the two stamped nuts on the pins. Use socket or pliers to press nuts on pins as far as they will go.

Step 25. Install wheel; remove jack stand; lower and remove the jack; and remove the wheel chocks. Proceed to next wheel and repeat steps 3 through 25.

Step 26. Refill master cylinder with new brake fluid (specify DOT-3) while assistant applies a gentle pumping pressure to the brake pedal.

Step 27. If calipers were overhauled, bleed brakes. (Refer to Chapter 12 for instructions.)

Step 28. Apply firm brake-pedal pressure several times (with engine running for power brakes) to seat pads against rotors.

Step 29. Test brakes. (Refer to Chapter 15 for instructions.) *CAUTION:* Do not move car until a firm brake pedal is obtained.

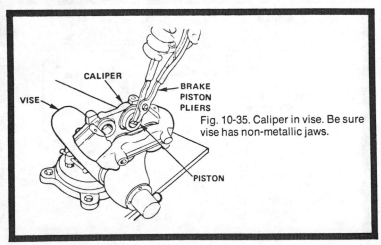

CALIPER

VISE

BRAKE
PISTON
PLIERS

Fig. 10-35. Caliper in vise. Be sure vise has non-metallic jaws.

PISTON

HOW TO OVERHAUL A DISC BRAKE CALIPER

Two basic caliper designs are used in American cars: the multipiston (two and four) caliper and the single-piston caliper. Because the similarities of all caliper designs outweigh the differences, only the two basic designs are treated separately in the step-by-step instructions presented below. And because hardly any special tools or materials are needed, no such list is provided; however, the items mentioned in the instructions include a vise, compressed air, a torque wrench, some ordinary wrenches, crocus cloth, brake fluid, and alcohol. But you should read the instructions appropriate for your job and decide for yourself whether or not you have adequate tools and materials at hand to tackle the task.

Multipiston Caliper

Step 1. Remove caliper from car to workbench.

Step 2. Open the bleeder screw and drain brake fluid from caliper.

Step 3. Remove crossover (transfer) tube (if any are present.)

Step 4. Remove caliper-bridge bolts; separate caliper halves.

Step 5. Remove any crossover seals.

Step 6. Position caliper in vise as shown in Fig. 10-35.

Step 7. Remove one boot at a time as shown in Figs. 10-36 and 10-37. Discard all rubber parts.

Step 8. Clean all reusable metal parts with alcohol or new brake fluid; blow through all passages to be sure they are open and clear. *CAUTION:* If compressed air is used, cover outlet with several folds of shop cloth as shown in Fig. 10-38 and apply pressure moderately. *NOTE:* Do not use a petroleum product (such as kerosene) to clean calipers—these products cause

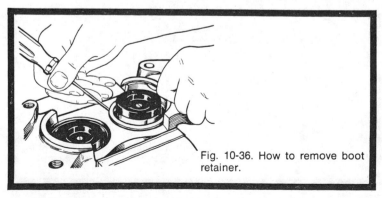

Fig. 10-36. How to remove boot retainer.

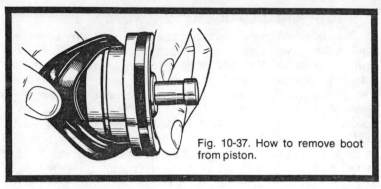

Fig. 10-37. How to remove boot from piston.

rubber parts to swell and are detrimental to brake pads.

Step 9. Inspect pistons for scratches and nicks. Use crocus cloth only to smooth piston walls. Clean after polishing.

Step 10. Inspect cylinder walls for scoring and deposits. Use crocus cloth to smooth cylinder walls. If cylinder wall does not clean up readily with crocus cloth, you may use a cylinder hone, as shown in Fig. 10-39. Cylinder bore must not be increased by more than 0.002 in. Use a medium- or fine-grit stone and a slow-speed drill. Clean cylinder and caliper.

Step 11. Inspect caliper for cracks and defects. Replace any badly worn or defective parts.

Step 12. Lubricate all new rubber parts to be used with clean new brake fluid.

Step 13. Assemble seals and boots on pistons.

Step 14. Lubricate cylinder bore with new brake fluid.

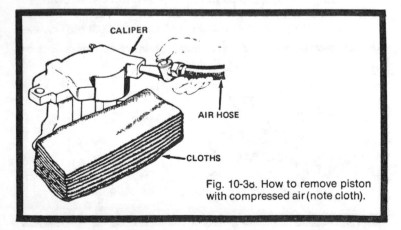

CALIPER

AIR HOSE

CLOTHS

Fig. 10-38. How to remove piston with compressed air (note cloth).

202

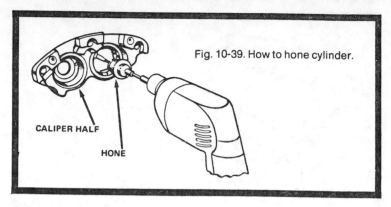

Fig. 10-39. How to hone cylinder.

CALIPER HALF

HONE

Step 15. Install seals on pistons. (This step applies to only a few early two-piston calipers.)

Step 16. Install piston springs behind pistons.

Step 17. Insert pistons into cylinders, open-end first. Avoid unseating or damaging any rubber parts.

Step 18. Bottom pistons in cylinders.

Step 19. On some models, seat dust boot as shown in Fig. 10-37.

Step 20. Close bleeder screw.

Step 21. Install crossover seals.

Step 22. Rejoin caliper halves. Insert and tighten caliper-bridge bolts as follows:

Brake Type	Torque (ft-lb)
Bendix (American Motors)	105
Bendix (all others)	130
Budd: small bolts (7/16 in.)	55
large bolts (5/8 in.)	150
Delco-Moraine	75
Kelsey-Hayes: Dart, Valiant, and Barracuda	75
Toronado	85

Step 23. Install transfer tube, if used.

Step 24. Install caliper.

Single-Piston (Floating) Caliper

Step 1. Remove caliper from car to workbench.

Step 2. Open the bleeder screw and drain brake fluid from caliper.

Step 3. Position caliper as shown in Fig. 10-40.

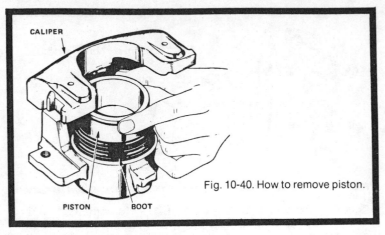

Fig. 10-40. How to remove piston.

CALIPER

PISTON   BOOT

Step 4. Remove piston from caliper cylinder. If compressed air is used, cover piston with several folds of shop cloth as shown in Fig. 10-41 and apply air slowly and cautiously. If piston does not pop out readily, remove air and tap caliper a few times with hammer. Discard all rubber parts.

Step 5. Clean all reusable metal parts with alcohol or brake fluid; blow through all passages to be sure they are open and clear. NOTE: Do not use a petroleum product (such as kerosene) to clean caliper—these products cause rubber parts to swell and are detrimental to brake pads.

Step 6. Inspect piston for scratches and nicks. Use crocus cloth only to smooth piston wall. Clean piston.

Step 7. Inspect cylinder wall for scoring or deposits. Use crocus cloth to smooth cylinder wall. Do not hone. Reclean entire caliper, especially cylinder.

Step 8. Inspect caliper for cracks and defects. Replace any badly worn or defective parts.

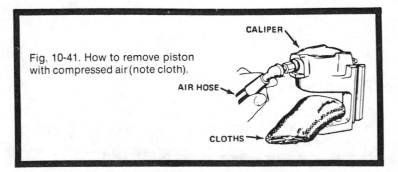

Fig. 10-41. How to remove piston with compressed air (note cloth).

CALIPER

AIR HOSE

CLOTHS

Fig. 10-42. How to assemble boot onto piston: Bendix and Budd.

Step 9. Lubricate all new rubber parts with new brake fluid.
Step 10. Lubricate piston and cylinder walls with new brake
fluid.

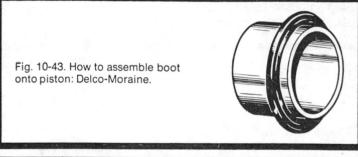

Fig. 10-43. How to assemble boot onto piston: Delco-Moraine.

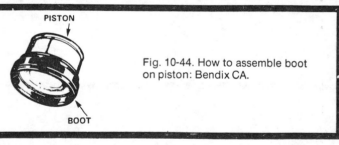

PISTON

Fig. 10-44. How to assemble boot on piston: Bendix CA.

BOOT

Step 11. Assemble new rubber parts on piston (Figs. 10-42 through 10-46).

Fig. 10-45. How to assemble boot on piston: Kelsey-Hayes.

PISTON

BOOT

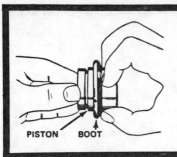

Fig. 10-46. How to assemble boot on piston: Vega.

PISTON   BOOT

Step 12. Insert piston into cylinder. Do not unseat rubber parts. Force piston into cylinder as far as it will go.

Step 13. On some models you must position piston boot into groove around cylinder (Fig. 10-47).

Step 14. Close bleeder screw.

Step 15. Install caliper.

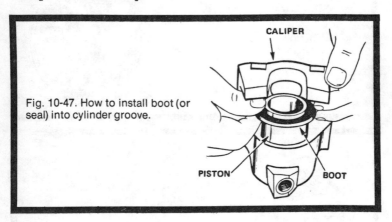

CALIPER

Fig. 10-47. How to install boot (or seal) into cylinder groove.

PISTON   BOOT

Step-by-Step Brake Line and Valve Repair

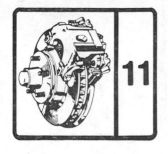

11

There are two varieties of brake lines: steel tubes and flexible hoses. The latter are used on wheel cylinders which must be free to move with the suspension.

METAL TUBES

Metal tubes can split; in the event of collision, they can be broken, kinked, or squeezed in such a way as to slow or block fluid flow. More often, however, they simply become loosened at the fitting. In the latter case, the only remedy required may be tightening the fitting.

FLEXIBLE HOSES

A flexible hose can split, rupture, weaken, fray, or collapse internally as well as become partially detached from a fitting. External flaws can usually be spotted by observing a hose for leakage or swelling while brake pressure is applied (with plenty of fluid in the hydraulic system). An internal flaw can only be spotted by removing a hose and looking through it toward a light—it should be open and have smooth walls.

TOOLS

Special tube wrenches (Fig. 11-1), properly fitted, are preferred for brake work; however, a snug-fitting open-end wrench or even an adjustable wrench is adequate for removing and replacing tube fittings. Be careful not to cross-thread fittings nor round off fittings with poorly sized wrenches.

The same information is applicable to hose fittings, which are less prone to cross-threading but more susceptible to damage by tearing, end fraying, and puncturing.

BRAKE INSPECTION

To examine brake tubes and hoses adequately you must visually inspect them (under heavy brake pressure—have an

Fig. 11-1. Special wrench for tube fittings (called flare-nut, tube-nut or tube-fitting wrench).

assistant press down hard on pedal) from master cylinder to all four wheels, for leaks, kinks, breaks, etc. Leaking fittings should be checked first for tightness. But, if tight and still leaking, remove the fitting and replace it. Any suspicious tube or hose should be replaced. Take the line or hose to be replaced to the parts store to be sure the replacement is of the correct length and size and has the proper fittings on both ends. If a tube must be bent 90° or more, use a tube bender.

Be sure the replacement tube or hose is tightened firmly with a wrench that fits snugly.

Following any tube or hose repair or replacement, refill the master cylinder and bleed the system as instructed in Chapter 12.

Examine repaired or replaced part for leakage.

Finally, test the brakes. (Refer to Chapter 15 for instructions on brake testing.)

VALVES AND BRAKE-WARNING LIGHT SWITCHES

Modern brake systems contain a brake-warning light and may contain one or more valves in addition to the residual-pressure check valve in the master cylinder.

You will need the following tools and supplies to replace valves:

- Flare-nut wrench.
- Open-end wrench of the proper size to remove valve-mounting nuts.
- Bleeder-screw wrench or substitute.
- Bleeder hose.
- Quart of DOT-3 brake fluid.

Metering Valve Replacement

The metering valve fits into the line somewhere below the master cylinder, and is usually attached to the engine wall. If you have determined that it is faulty, it must be replaced.

CAUTION: If the brake-warning light remains lit, refer to instructions below before bleeding the system.

Step 1. Using a flare-nut wrench, disconnect both hydraulic lines from valve.

Step 2. Detach metering-valve assembly from car. Discard old valve after you have made sure new valve is identical to it.

Step 3. Connect both hydraulic lines to new valve; be sure valve is properly positioned for installation.

Step 4. Install valve on engine-compartment wall.

Step 5. Tighten flare nuts.

Step 6. If you are servicing the brake system on an American Motors, International, or Kaiser-Jeep vehicle, refer to the appropriate section below for instructions on how to avoid damage to the brake-warning light switch.

Step 7. Bleed brake system as required (Chapter 12).

Step 8. Test brakes as required (Chapter 15).

Proportioning Valve Replacement

The proportioning valve (Fig. 11-2) fits into the hydraulic line somewhere below the master cylinder; it is usually attached to the engine compartment wall by screws or stud bolts. If you have determined that it is faulty, then it must be replaced. *CAUTION*: If the brake-warning light has come on and remains on, refer to the section on this switch below before bleeding air from system.

Step 1. Use a flare-nut wrench and disconnect all hydraulic lines from valve.

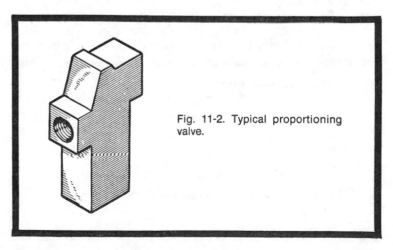

Fig. 11-2. Typical proportioning valve.

Step 2. Detach proportioning valve assembly from engine wall. Discard old valve *after* you have made sure new valve is identical to it in appearance and function.

Step 3. Connect all hydraulic lines to new valve; be sure valve is properly positioned for installation.

Step 4. Install valve onto engine compartment wall.

Step 5. Tighten flare nuts.

Step 6. If you are servicing the brake system on an American Motors, International, or Kaiser-Jeep vehicle, see below for instructions on how to avoid damage to brake-warning light switch. For other makes, proceed to step 7.

Step 7. Bleed brake system as required (Chapter 12).

Step 8. Test brakes as required (Chapter 15).

Combination Valve Replacement

The combination valve (Fig. 11-3) fits into the hydraulic lines somewhere below the master cylinder, within the engine compartment, and is usually attached to the left-hand engine wall by stud bolts. If you have determined that it is faulty, then it must be replaced.

Step 1. Use a flare-nut wrench and disconnect the several hydraulic lines from valve.

Step 2. Detach combination valve from engine wall. Discard old valve *after* you have made sure the new one is the *exact* replacement.

Step 3. Connect all the hydraulic lines to the new valve. Draw fittings up snugly, but do not tighten.

Step 4. Install valve onto engine compartment wall.

Step 5 Tighten flare nuts.

Step 6. If you are servicing the brake system on an American Motors, International, or Kaiser-Jeep vehicle, and the brake-warning light switch is not a

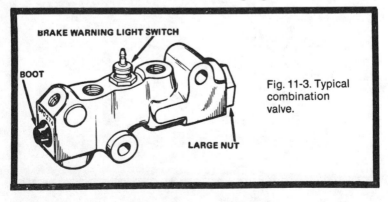

BRAKE WARNING LIGHT SWITCH

BOOT

LARGE NUT

Fig. 11-3. Typical combination valve.

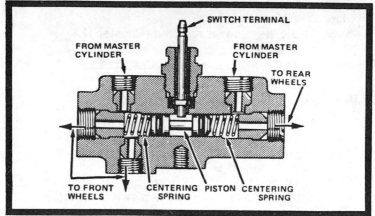

Fig. 11-4. Cross section of typical brake-warning light switch (single piston with centering springs).

part of the combination valve, refer to the appropriate section below for instructions on how to avoid damage to the brake-warning light switch before proceeding to step 7. Otherwise proceed to step 7.

Step 7. Bleed brake system as required. (Chapter 12.)

Step 8. Test brakes as required (Chapter 15.)

AMERICAN MOTORS, INTERNATIONAL, KAISER-JEEP BRAKE-WARNING LIGHT SWITCH REPLACEMENT

You will need these tools and supplies:

- Flare-nut wrench or substitute.
- Open-end wrench (or other tool) of the proper size to remove switch-mounting nuts, bolts, screws.
- Bleeder-screw wrench or substitute.
- Bleeder hose.
- Small can of DOT-3 brake fluid.
- Small empty container.

Single-Piston Switch with Centering Springs

This variant is shown in Fig. 11-4 in cross section.

Step 1. Disconnect the electrical lead from the switch.

Step 2. Use a flare-nut wrench and disconnect all hydraulic lines from switch.

Step 3. Detach switch from mounting. Discard switch *after* you have made sure new switch is identical.

Step 4. Connect hydraulic lines to switch; pull flare nuts up snugly, but do not tighten.

Step 5. Install switch on mounting.

Step 6. Tighten flare nuts.

Step 7. Bleed system as required (Chapter 12.)

Step 8. Connect electrical wire to switch.

Step 9. Test brakes as required (Chapter 15.)

Two-Piston Switch with Centering Springs

This variant is shown in Fig. 11-5.

Step 1. Disconnect the electrical lead from the switch.

Step 2. Use a flare-nut wrench and disconnect all hydraulic lines from switch.

Step 3. Detach switch from mounting. Discard old switch *after* you have made sure new switch is identical.

Step 4. Connect hydraulic lines to switch; pull flare nuts up snugly, but do not tighten them.

Step 5. Install switch to mounting.

Step 6. Tighten flare nuts.

Step 7. Remove switch terminal from switch.

Step 8. Bleed system as required (Chapter 12.)

Step 9. Install switch terminal into switch.

Step 10. Connect electrical wire to switch terminal.

Step 11. Test brakes as required. (Refer to Chapter 15 for instructions.)

Two-Piston Switch without Centering Springs

This variant is shown in Fig. 11-6.

Step 1. Disconnect the electrical lead from switch.

Step 2. Use a flare-nut wrench and disconnect all hydraulic lines from switch.

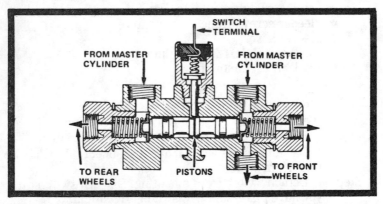

Fig. 11-5. Cross section of typical brake-warning light switch (two pistons, with centering springs).

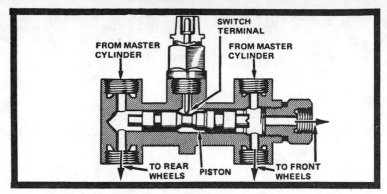

Fig. 11-6. Cross section of typical brake-warning light switch (two pistons, without centering springs).

Step 3. Detach switch from mounting. Discard old switch *after* you have made sure new switch is identical to old one in apperance and function.

Step 4. Connect hydraulic lines to switch; pull flare nuts up snugly, but do not tighten them.

Step 5. Install switch to mounting.

Step 6. Tighten flare nuts.

Step 7. Bleed system as required (Chapter 15).

Step 8. Connect electrical wire to switch terminal.

Step 9. If brake-warning light comes on when wire is attached and the ignition switch is turned *on*, then the brake-warning light piston must be centered as follows:

(a) Attach bleeder hose to a bleeder screw in the system (front or back) opposite to the one bled last; place other end of hose in container partially filled with new brake fluid.

(b) Turn ignition switch *on*.

(c) Open bleeder screw slightly as an assistant presses gently on the brake pedal. Tighten bleeder screw the instant brake-warning light goes out and while assistant maintains pedal pressure.
NOTE: If brake warning light flickers off and then comes on again, the piston has moved too far. The entire procedure must be repeated with hose attached to opposite system.

(d) When piston is centered, turn ignition switch *off*, detach bleeder hose, and proceed to step 10.

Step 10. Test brakes as required. (Refer to Chapter 15 for instructions.)

Step-by-Step Bleeding and Flushing

12

The reasons for flushing and bleeding a brake system are enumerated in Chapter 1. In this section we deal specifically with the tools and materials you must have in hand and the steps to be performed to bleed and flush your system properly and with a minimum of difficulty.

WHAT YOU NEED

The tools and supplies that you need are relatively inexpensive and few in number.

Flushing the System

Simple flushing is accomplished with brake fluid only; however, if the system has become contaminated with a petroleum product, rust, or other foreign matter, it should be flushed with denatured alcohol until all old fluid and contaminants are removed. To accomplish a thorough flushing, you need these items (illustrated in Fig. 12-1):

- Quart of DOT-3 brake fluid.
- Quart of denatured alcohol (purchase at drugstore).
- Bleeder hose.
- Container to catch fluid (preferably a glass jar).
- Wrench to fit bleeder screws. A special wrench is made for this purpose, an offset box-end that comes in three popular sizes: $\frac{1}{4}$ in., $\frac{5}{16}$ in., and $\frac{3}{8}$ in.

Bench-Bleeding the Master Cylinder

It is certainly practical, if not essential, to bench bleed any repaired or replacement master cylinder before you install it

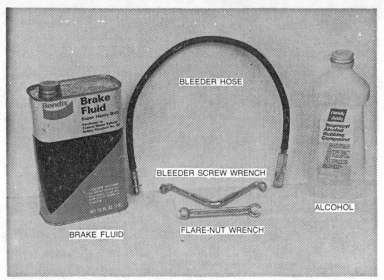

Fig. 12-1. Tools and supplies needed to flush hydraulic system.

on your car. The idea is to remove the large quantity of air from the master-cylinder bore and thereby save time and brake fluid. To bench-bleed your system, you need the following items:

- Can of DOT-3 brake fluid.
- Bleeder tube (one for each reservoir) fitted for your master cylinder (Fig. 12-2).
- Outlet-port plugs (Fig. 12-3).

Bleeding the Wheel

Depending upon where your hydraulic system is opened to the air, bleeding may be necessary at only one wheel or at all four wheels, and possibly at the master cylinder as well.

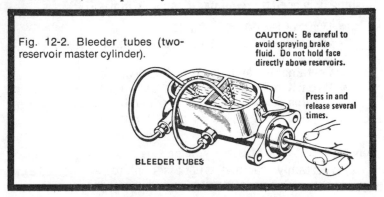

Fig. 12-2. Bleeder tubes (two-reservoir master cylinder).

CAUTION: Be careful to avoid spraying brake fluid. Do not hold face directly above reservoirs.

Press in and release several times.

BLEEDER TUBES

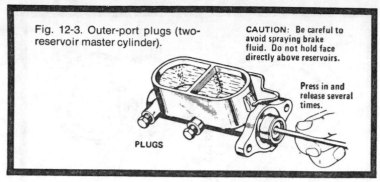

Fig. 12-3. Outer-port plugs (two-reservoir master cylinder).

CAUTION: Be careful to avoid spraying brake fluid. Do not hold face directly above reservoirs.

Press in and release several times.

PLUGS

Typical bleeding points are shown in Fig. 12-4. In any event, the following items are needed to wheel-bleed your brake system:

- Can of DOT-3 brake fluid. NOTE: If system is low on fluid, you may need more than a single can of fluid, especially if you plan to bleed system at all four wheels.
- Bleeder hose.
- Container to catch fluid (preferably a glass jar).
- Wrench to fit bleeder screws. A special wrench is made for this purpose, an offset box-end type that comes in three popular sizes: $^1/_4$ in., $^5/_{16}$ in., and $^3/_{16}$ in.

There are two methods of bleeding a hydraulic system, both of which are discussed below.

HOW YOU DO IT

Flushing and bleeding, the last operations in a repair job, are quite critical.

Flushing a Hydraulic System

Step 1. Determine what type of brake-warning light switch your car has (refer to Table 3-1). If it has the two-piston type with centering springs—the type used on some American Motors, International, and Kaiser-Jeep vehicles—you must remove the terminal plug from the switch to prevent damage to the unit.

Step 2. Fill master cylinder to mark with appropriate fluid—either brake fluid or denatured alcohol. The fluid level should be ¼ in. or so from the top of the reservoir.

Step 3. Attach bleeder hose to bleeder screw of right rear wheel; insert other end of hose into glass jar (Fig. 12-5). Open bleeder screw.

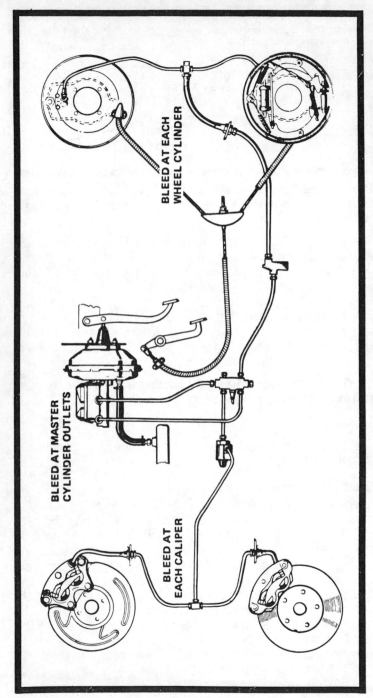

BLEED AT EACH WHEEL CYLINDER

BLEED AT MASTER CYLINDER OUTLETS

BLEED AT EACH CALIPER

Fig. 12-4. Bleeding points for any hydraulic brake system.

218

Step 4. Have an assistant pump brake pedal gently until fluid no longer flows from bleeder screw. Close screw.

Step 5. Refill master cylinder.

Step 6. Repeat steps 3 and 4 at each wheel and step 5 after each except the last.

Step 7. Without refilling master cylinder, bleed system at each wheel until no fluid flows from any.

Step 8. Refill master cylinder with brake fluid.

Step 9. Bleed each wheel individually once more, this time with jar partially filled with new brake fluid and end of bleeder hose submerged to prevent sucking air back into the system when pedal pressure is released. When air bubbles no longer appear, close bleeder screw. Refill master cylinder and repeat bleeding procedure at each wheel.

Step 10. Refill master cylinder.

Step 11. Repeat step 9 at bleeder screw on combination valve, if your car has one.

Step 12. Refill master cylinder.

Step 13. Repeat step 9 at bleeder screw on master cylinder, if your master cylinder has one, or by loosening the outlet-tube nuts and depressing brake pedal one time, slowly and gently, and tightening nuts while pressure is still being applied.

Step 14. Refill master cylinder; install master-cylinder cover.

Step 15. Test your brakes in accordance with the instructions given in Chapter 15.

Bench-Bleeding a Master Cylinder

Step 1. Install bleeder tubes or plugs.

Step 2. Fill master cylinder with new brake fluid.

Step 3. Gently press in and release master cylinder piston several times.

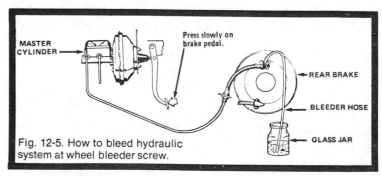

Fig. 12-5. How to bleed hydraulic system at wheel bleeder screw.

Step 4. Remove bleeder tubes or plugs. Don't be concerned if some fluid runs out through outlet ports.

Step 5. Install cover on master cylinder.

Step 6. Install master cylinder on car. Be sure hydraulic lines are firmly connected.

Step 7. Remove master cylinder cover and refill master cylinder.

Step 8. Install bleeder hose on bleeder screw at combination valve, if your car has one.

Step 9. Submerge other end of bleeder hose in glass jar partially filled with new brake fluid.

Step 10. Have assistant press brake pedal gently until no air bubbles appear in fluid in jar; then, with slight pressure still applied to brake pedal, close bleeder screw.

Step 11. Repeat steps 8, 9, and 10 with hose attached to bleeder screw on wheel cylinder at each wheel, one at a time. Refill master cylinder after each bleeder screw is closed.

Step 12. Be sure master cylinder is filled with fluid, then install cover on master cylinder.

Step 13. Test your brakes in accordance with the instructions in Chapter 15.

Bleeding the Wheel

Step 1. Fill master cylinder to mark with new brake fluid. The fluid should be ¼ in. or so from the top of the reservoir.

Step 2. Install hose on bleeder screw of wheel cylinder (Fig. 12-5) and submerge other end of hose in partially filled jar of brake fluid.

Step 3. Have assistant apply slight pressure to brake pedal as you open bleeder screw.

Step 4. Continue brake pedal pressure until bubble-free fluid flwos from hose (pump pedal more than once if necessary); then, with slight pressure still applied, close bleeder screw.

   NOTE: If more than one wheel is to be bled, begin with the one farthest from master cylinder and finish with one nearest master cylinder.

   NOTE: Repeat steps 1, 2, 3, and 4 for each wheel to be bled.

Step 5. Refill master cylinder, and install cover on it.

Step 6. Test your brakes in accordance with the instructions in Chapter 15.

Step-by-Step Parking Brake Adjustment and Overhaul

13

Most parking brakes on American cars are of one general type, the *integral* brake, so-called because it is an integral part of the rear-wheel drum brake. There are two other types that have been used since World War II, and each is given consideration below.

WHAT YOU NEED

Tools and supplies vary a bit between brake types and are discussed seprately.

Integral Parking Brakes

For items needed to service those parts found inside the brake drums, refer to the discussion of drum brakes in Chapter 9.

To perform routine service on the parking-brake cable system (Fig. 13-1), you need the following items:

- Pair of pliers (if cable must be removed).
- Small squirt can of penetrating oil.
- Box-end or socket wrench of proper size (if cable must be removed).

One nonintegral parking brake—that of the Corvette—is covered here because the step-by-step instructions for serviing this brake are identical with those given for the integral type. Similarly, the procedures are limited to servicing the linkage external to the drum. Figure 13-2 illustrates the Corvette parking brake.

Transmission-Type Parking Brake

Like the Corvette brake mentioned above, these nonintegal parking brakes (Fig. 13-3) aren't likely to be worn

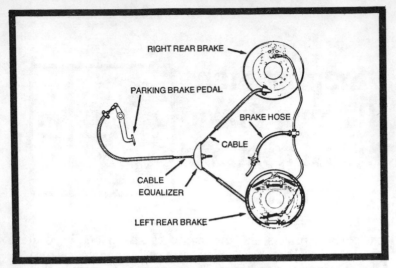

Fig. 13-1. Integral parking brake.

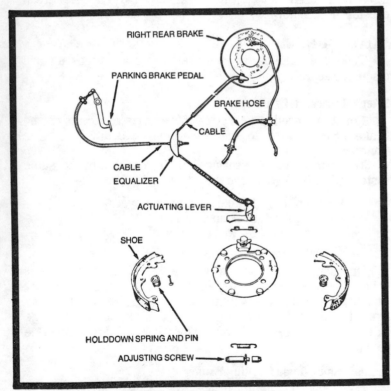

Fig. 13-2. Corvette parking brake.

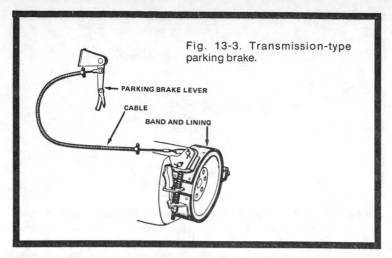

Fig. 13-3. Transmission-type parking brake.

PARKING BRAKE LEVER

CABLE

BAND AND LINING

out by normal (parking) use. Their major weakness is that drivers too often forget to release them when the car is put into motion. In addition to listing those items needed to service the cable linkage external to the brake, adjusting needs are also listed:

- Small can of penetrating oil.
- Pair of pliers (if cable must be removed).
- Box-end or socket wrench of proper size (if cable must be removed).
- Box-end or open-end wrenches of proper size to adjust external-band parking brake.
- Medium-blade screwdiver to adjust internal-band parking brake.

ADJUSTING CABLE SYSTEMS

Each brake type requires a slightly different technique.

Integral Parking Brakes

Step 1. Rear-wheel brakes must be operative for this parking brake to work adequately; therefore, any needed rear-wheel brake service sould be completed before attempting to adjust parking brake.

Step 2. Check parking-brake cables for free movement in conduits. If necessary, squirt penetrating oil into conduits to free cables. (Refer to Fig. 13-4.)

Step 3. Check for worn equalizer or other linkage parts.

Step 4. Check cable for broken strands.

223

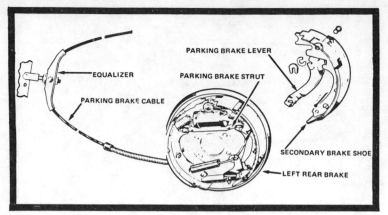

Fig. 13-4. Service points of a typical integral parking-brake cable system.

Step 5. If all parts are in good working condition, set parking brake lever at about two-thirds of its normal travel.(If parts are not in good condition, replace them before continuing, beginning with this step.)

Step 6. Tighten adjusting screw behind equalizer until brakes are set lightly but firmly.

Step 7. If rear cable is not properly positioned in equalizer, rear-wheel brakes cannot provide equal braking action, so be sure this cable is properly positioned.

Corvette Parking Brakes

Step 1. Parking-brake shoes press against inside of disc-brake rotor on each rear wheel (Fig. 13-5). To adjust, set parking-brake pedal (or lever) to first notch; then rotate adjusting screw at each wheel until shoes are firm against rotor drum. Release

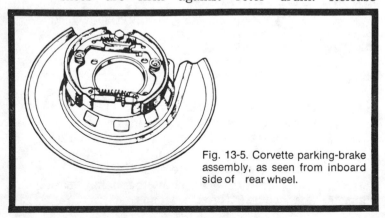

Fig. 13-5. Corvette parking-brake assembly, as seen from inboard side of rear wheel.

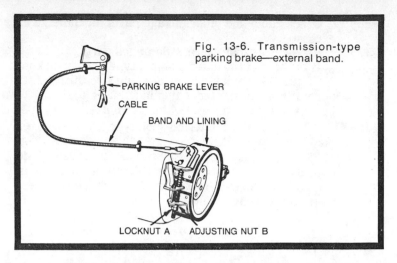

Fig. 13-6. Transmission-type parking brake—external band.

PARKING BRAKE LEVER

CABLE

BAND AND LINING

LOCKNUT A   ADJUSTING NUT B

parking-brake pedal (or lever). There should be no drag on rear wheel.

Step 2. Inspect parking-brake linkage, cable, and equalizer for worn, frayed, or damaged parts. Repair or replace parts as necessary.

Step 3. Check parking-brake cable for free movement in conduits. If necessary, squirt penetrating oil into conduits to free cable.

Step 4. Be sure cables are properly positioned in equalizer to apply equal pressure to both rear wheels.

Transmission-Type Parking Brakes

Step 1. Check parking-brake cable for free movement in conduit. If necessary, squirt penetrating oil into conduit to free cable.

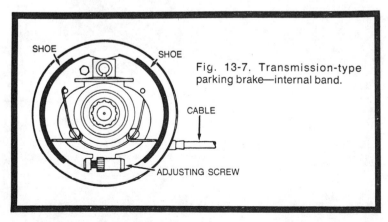

SHOE   SHOE

Fig. 13-7. Transmission-type parking brake—internal band.

CABLE

ADJUSTING SCREW

Step 2. Check cable and other linkage parts for damage or excessive wear.

Step 3. Replace or repair worn or damaged parts.

Step 4a. If brake is of the external-band type, loosen locknut A (Fig. 13-6); set hand lever at about two-thirds of its normal travel; tighten adjusting nut B until brake band is snug upon drum; tighten locknut A.

Step 4b. If brake is of the internal-shoe type, remove cover plate from back of brake; rotate adjusting screw (Fig. 13-7) by pushing against it with screwdriver blade until shoe is snug against drum—with brake lever in released position—then back it off about one complete turn. Replace cover plate

Step-by-Step Lamp and Switch Repair

14

There are no more than three or four lamps in the brake system: two in the stoplight system and one or two to indicate the condition of parking and regular brakes. And servicing these lamps is quite simple—the most common failure being a burned-out bulb and the second most common being a blown fuse. Thus, our tool list for brake-light servicing is practically nil, since none is needed for fuse replacement and, often as not, none is needed for bulb replacement. If an item (either a tool or part) is needed, it is mentioned in the step-by-step instructions.

Step 1. Check the fuse first. To locate fuse block, refer to Table 14-1 or look around behind the instrument panel. Usually the fuse block is mounted on the firewall somewhere near the steering column, but it might be on the firewall, under the dash on the passenger side (near glovebox), or even under the hood. A typical fuse block is shown in Fig. 14-1. Your owner's manual may indicate which fuse protects the various bulbs in the system. If it doesn't, check them all—it only takes a minute. Remove the fuse, hold it up to light, and determine whether or not fuse is good (Fig. 14-2). A bad fuse is one in which the filament has a break, however small.

Step 2. If fuse is burned out, replace it with new one of proper size and rating (amperage is stamped or printed on metal end piece).

Step 3. If fuse is okay check the bulb. Access to stoplight bulb may be through your car's trunk, as shown in Fig. 14-3. Or you may have to remove taillight lens with a

Table 14-1. Fuse Block Location Guide.

AMERICAN MOTORS: Beneath dash (instrument panel), on passenger side, next to parking brake mechanism.

CHRYSLER:

Imperial	Beneath instrument panel on driver side.
Challenger	Beneath instrument panel on driver side.
Coronet	Beneath instrument panel on driver side.
Charger	Beneath instrument panel on driver side.
Cordoba	Beneath instrument panel on driver side.
Dart	On lower instrument-panel reinforcement strut.
Polara	Beneath instrument panel on driver side.
Monaco	Beneath instrument panel on driver side.
Valiant	On lower edge of instrument panel.
Duster	On lower edge of instrument panel.
Baracuda	On cowl under instrument panel on driver side.
Satellite	Beneath instrument panel on driver side.
Fury	Beneath instrument panel on driver side.
Grand Fury	Beneath instrument panel on driver side.

FORD:

Thunderbird	On dash panel, left of steering column.
Mark III & IV	On dash panel, left of steering column.
Continental	Under instrument panel, above parking-brake mechanism.
Ford	Under instrument panel on driver side.
Mercury	Under instrument panel on driver side.
Meteor	Under instrument panel on driver side.
Mustang	On parking-brake support flange.
Mustang	On parking-brake support flange.
Cougar	On parking-brake support flange.
Maverick	On instrument panel, left of steering column.
Comet	On instrument panel, left of steering column.
Torino	On instrument panel, left of steering column.
Montego	On instrument panel, left of steering column.
Pinto	On instrument panel, left of steering column.
Granada	On instrument panel, left of steering column.
Monarch	On instrument panel, left of steering column.

Table 14-1. Cont'd.

GENERAL MOTORS:

Buick	On driver side, beneath instrument panel, on cowl.
Apollo	On driver side, beneath instrument panel, on cowl.
Nova	On driver side, beneath instrument panel, on cowl.
Ventura	On driver side, beneath instrument panel, on cowl.
Omega	On driver side, beneath instrument panel, on cowl.
Astre	On driver side, beneath instrument panel.
Monza	On driver side, beneath instrument panel.
Skyhawk	On driver side, beneath instrument panel.
Starfire	On driver side, beneath instrument panel.
Vega	On driver side, beneath instrument panel.
Cadillac	On driver side, beneath instrument panel, on cowl.
Chevrolet	On driver side, beneath instrument panel.
Oldsmobile	Beneath instrument panel, on cowl.
Pontiac	Beneath instrument panel, above floor mat, on dash shroud.

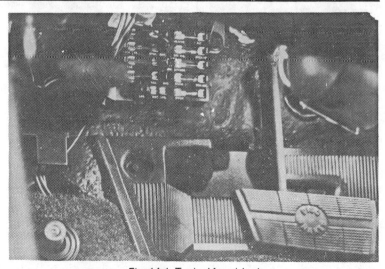

Fig. 14-1. Typical fuse block.

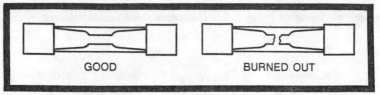

Fig. 14-2. How to determine condition of a fuse.

Phillips screwdriver, as shown in Fig. 14-1. For those removed from inside the car, rotate bulb holder and pull it out. For either type, to remove bulb, push in, and rotate it counterclockwise. Inspect bulb; it should have an unbroken filament.

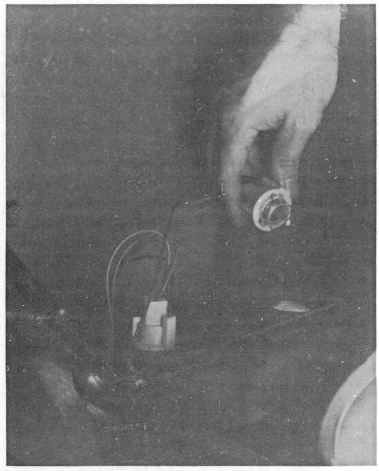

Fig. 14-3. Brake light access from inside car trunk.

Fig. 14-4. Brake light access by removing taillight lens.

Step 4. If bulb is burned out, replace it with new one with same base and rating. Install bulb into holder, and holder into reflector or lens over light.

Step 5. If bulbs are okay, check wiring. Be sure all connections are good and wire is neither broken nor rubbed bare and shorted against some metal part.

Step 6. If wire is defective, repair or replace it.

Step 7. If wire is okay, check stoplight switch. Be sure it is positioned properly and mounted firmly—button should be depressed when brake pedal is pressed about ½ in.

Step 8. If switch is positioned properly but light still doesn't come on, use a short piece of automotive electrical wire to "jump" switch. Connect the wire between the two switch terminals.

Step 9. If stoplights come on, switch is defective. Throw it away and install a new one. If light does not come on, seek the services of a qualified automotive electrician.

Step-by-Step Brake System Tests

15

After any brake service work, your brakes should be tentatively tested *before* they are road-tested.

Step 1. With engine off, press and release brake pedal several times. Then hold pedal depressed lightly and start engine. Pedal should drop slightly when engine starts and then hold firm.

Step 2. Apply heavy foot pressure to pedal (with engine running for power brakes) and check for sponginess and pedal reserve. Pedal should feel firm and foot pad should be a minimum of 2 in. from the floor for manual brakes and 1 in. for power brakes.

Step 3. Hold pedal depressed with medium foot pressure for about 15 seconds. Pedal should not drop under this steady pressure.

Step 4. For power brakes only: run engine to medium speed and shut down. Wait about 90 seconds and apply brakes several times. The first two or three applications should be power-assisted, then pedal should get firmer (indicating no power assist).

Step 5. For automatic-adjuster equipped brakes only: make four or five forward and reverse stops to allow automatic adjusters to function. Brake should begin to feel right.

Step 6. Make several low-speed stops, each from a slightly increased speed, to test braking effectiveness. If brakes appear to be working properly, proceed to road-test. Test brakes by stopping car from progressively higher speeds (for example. 35, 45, and 55 mph.)

Index

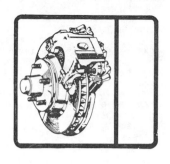

General Motors Corporation—Bendix drum

Bendix duoservo drum brakes are found on all GM passenger cars rear and front (as applicable). Vega nonservo rear brakes are the sole exception.

General Motors Corporation—Bendix disc (front only)

Buick (Electra, Invicta, LeSabre, Rivera, Wildcat)—1967—69

General Motors Corporation—Delco-Moraine disc (front only—except Corvette)

Buick—1970—74
Cadillac—1968—74
Chevelle—1967—74
Cheverolet—1967—74
Corvette—1965—74
Eldorado—1968—74
F-85 and Cutlass—1967—74
Firebird—1967—74
Grand Prix—1968—74
Monte Carlo—1970—74
Nova—1967—74
Oldsmobile—1967—71
Pontiac—1967—74
Riviera—1970—74
Special (Buick)—1967—74
Tempest (includes LeMans and GTO)—1967—74
Toronado—1969—74
Vega—1971—74
Ventura—1971—74